Basil Ede's
BIRDS

Basil Ede's BIRDS

With text by

Robert Dougall

Vice-President RSPB

Foreword by

HRH The Prince Philip
Duke of Edinburgh

SEVERN
HOUSE

This book is published in association with
Royle Publications Limited

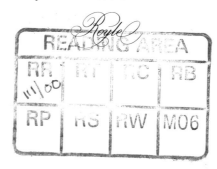

British Library Cataloguing in Publication Data
Ede, Basil
Basil Ede's Birds
1. Ede, Basil
2. Birds in art
3. Birds – Europe
I. Dougall, Robert
759.2   ND497.E3/
ISBN 0-7278-2005-2

Published by
Severn House Publishers Limited
144–146 New Bond Street London W1Y 9FD
Plates and introduction © Basil Ede 1966–1979
Text accompanying plates © Robert Dougall 1980
Editorial Ian Jackson
Designer Michael Stannard

ISBN 0 7278 2005 2

Phototypeset by Filmtype Services Limited, Scarborough
Printed and bound by New Interlitho SpA, Milan, Italy

BUCKINGHAM PALACE

     If you cannot spend as much time bird-watching
as you would like - and that applies to every
enthusiast - the next best thing is to revive your
memories of birds watched by looking at Basil Ede's
pictures. This book is a welcome sequel to his
'Birds of Town and Village' and between them they
provide a wonderful selection for the delectation
of the armchair bird-watcher. For those who can
tear their eyes away from the pictures the text by
Robert Dougall is an additional pleasure.

1980

Preface

This book which contains full colour reproductions of seventy birds of the British Isles and northern Europe, painted by Basil Ede, will be a cherished source of delight and interest to its owners wherever they may live. It will also contribute greatly to bird welfare, for pictures of birds, especially those faithfully portraying them in their natural settings, in all their grace, charm and loveliness, create a greater appreciation of them than the finest verbal descriptions.

I first saw the bird paintings of Basil Ede in 1966 at the Kennedy Galleries in New York City. I remember being greatly impressed with his work and in particular with his extraordinary ability to delineate and depict feathers. Receiving an invitation thirteen years later in 1979 to an exhibition of his recent paintings, and at the very same galleries, I eagerly accepted. Again I was struck, and this time more forcefully than before, by the strikingly lifelike quality Ede imparts to his subjects. They were all alive, beautiful and exciting. I stood long before one of the smallest of the paintings, that of the Barn Swallow at the nest. One of the parent birds was presented as having just alighted, with wings still partially uplifted and the deeply forked tail spread wide. I marvelled that the motion of alighting had been stopped as if by high speed photography and yet the sense of continuing movement remained, the illusion that those wings were still closing! I noted with admiration that the bird was authentically a Barn Swallow in every detail of its anatomy, in all its colour and remarkably so in its *personality*. All the qualities I appreciated in this little canvas are to be seen in every Ede painting.

Technical ability is, to be sure, essential for the artist, but to impart lifelike qualities and personality to one's subjects requires the abilities of the true master. Yet there is something else needed: a knowledge of the bird itself, of its haunts and habits. This can be acquired only from long hours in the field, in an intimate study of the living bird, from field notes, sketches and even photographs. Every Ede painting gives eloquent manifestation of such knowledge and understanding.

I stated earlier that this book will be of value to the birds themselves. Indeed it will be a boon for no one can behold these beautiful paintings and read the excellent descriptive texts without being moved to a greater appreciation of birds. Bird painting has contributed enormously to awakening and forwarding an interest in and concern for birds. Through the paintings of the birds he loves Basil Ede has become a valued contributor to their preservation.

Carl W. Buchheister

Carl W. Buchheister
President Emeritus
National Audubon Society

Bethesda
Maryland 1980

Contents

Foreword by HRH The Prince Philip, Duke of Edinburgh
Preface by Carl W. Buchheister
Introduction by Basil Ede

The plates

Blue Tit	19	Tufted Duck	73
Hawfinch	20	Redstart	74
Bullfinch	21	Spotted Flycatcher	75
Chaffinch	23	Osprey	77
Little Owl	25	Green Woodpecker	79
Cuckoo	27	Skylark	80
Blackcap	28	Yellowhammer	81
Goldfinch	29	Lesser Spotted Woodpecker	83
Pied Flycatcher	31	Woodcock	85
Willow Warbler and Chiffchaff	33	Coal Tit	86
Greenfinch	35	Long-tailed Tit	87
Mallard	36	House Martin	89
Canada Goose	37	Blackbird	91
Tree Sparrow and House Sparrow	39	Kestrel	93
Lapwing	41	Treecreeper	94
Kingfisher	43	Bluethroat	95
Puffin	45	Pheasant	97
Jay	46	Tawny Owl	99
Barn Owl	47	Pintail	101
Mute Swan	49	Nuthatch	103
Pied Wagtail	50	Wren	105
Grey Wagtail	51	Fieldfare	106
Great Crested Grebe	53	Redwing	107
Hoopoe	55	White-fronted Goose	109
Great Tit	57	Siskin	110
Whitethroat	58	Waxwing	111
Nightingale	59	Goldeneye	113
Great Spotted Woodpecker	61	Goldcrest	114
Avocet	63	Snipe	115
Mistle Thrush	64	Teal	117
Song Thrush	65	Firecrest	119
Red Grouse	67	Robin	121
Bee-eater	68	Grey Heron	123
Golden Oriole	69	Bearded Tit	125
Swallow	71	Black Redstart	127

Introduction

The paintings collected together in this book span nearly fifteen years of my life as a bird painter. They represent many of the most familiar and attractive birds to be seen regularly in Britain and northern Europe. Some are resident all year round, others visit us in winter or summer.

The original paintings, from which the plates were made, are now scattered far and wide, but to look at these superb reproductions of them recalls for me many enjoyable memories and experiences. The titmice, with their alert and agile movements, were my 'nursery' subjects before graduating to larger birds such as the Pheasants which exploded from under my feet during country walks, or ducks which flew in fast arrowhead formation through the dusk. Then there are the Swallows and martins, harbingers of spring, whose arrival brings cheerful anticipation of the summer to come. Getting to know these and many others, as they come and go through the seasons, has added a great deal of enjoyment to my life.

I was brought up in the Surrey countryside near Guildford and, as a boy, spent long hours roaming the fields and woods. Such places were full of interest, providing trees to climb, undergrowth to use as hide-outs and sunny glades where I could daydream contentedly. To me this was blissful contrast to the school classroom where I ran the constant risk of having my hair pulled out or a caning. The woods held no such terrors.

Certainly I never went out with the intention of studying wild birds and animals, but I became very much aware of their presence as I indulged in my favourite pastime of tracking an imaginary quarry through the trees. Either alone or with friends, I was learning the art of stalking quietly as a family of young kittens or Fox cubs might do learning to hunt. A whole new world was revealed to me where ducks reared platoons of downy ducklings on a woodland pond and owls, with soft, subdued feathers, filled the interiors of large tree holes and looked at me with eyes that told me not to go too close. I had perhaps a face to face encounter with a Red Squirrel (unhappily no longer resident in most of our woodland) or a glimpse of Fallow Deer prancing noisily through the trees with tails held high. Less happily, I witnessed the drama of a Fox pouncing on a Pheasant only a metre or so away as dusk turned the forest into gloom. It was a source of amazement to me that so much was going on behind those leafy backdrops whilst I made my own use of the territory.

Notwithstanding my long hours in the forests, as a nine-year-old boy, I drew endless pictures of the most significant aspect of my environment. It was wartime and, although I could not fully understand what had caused whole nations to harbour such far-reaching hatred, I was fascinated by the machines of war which were beginning to appear on the ground and in the air. I became as much attracted to planes, tanks and guns as my elders had been to steam engines. The Spitfire fighters, which flung themselves about the sky, were irresistible subjects to draw. So too were the slim, pencil shapes of the visiting Dornier 17 bombers with pale undersides and sinister black crosses. I drew many pictures of these and other famous aircraft of World War II.

During the war years I even interested myself in designing aircraft of my own. Detailed plans and projections were laboriously drawn out with inspiration coming from the skies and the many aviation periodicals I bought with my pocket money. Sir Barnes Wallis, inventor of the Bouncing Bomb, lived nearby, and one evening, when he was visiting our house he turned his attention to one of my designs. Given some technical improvements he thought it would fly and so set about showing me how to make the alterations. In the absence of proper drawing instruments Sir Barnes made use of pins and pieces of cotton. Although I was only eleven, the tall white-haired man with penetrating but kind eyes talked to me at length about aerodynamics as if to a colleague. Despite this treasured master class I was never to become an aircraft designer.

If man-made machines appealed to my restless pencil, so too did people themselves. Observing and caricaturing the meaningful lines in a face became an obsession. My 'victims' were usually those in authority and it was immensely satisfying to have such a weapon at my disposal. I could undermine or ridicule at will, at the same time earning the respect of my schoolmates. I drew on any surface readily to hand, be it school exercise book or textbook. I could not stop myself because there was always someone or something to be drawn.

About this time I became acquainted with L.R. Brightwell, author, eminent natural history artist, cartoonist and an artist for the Zoological Society of which he was a Fellow. His house was a source of endless fascination to me but dismay to many others. It housed a huge collection of skeletons and skins of wild animals which had died and been sent to him for study. Each bone had been cleaned and carefully reconstructed. Mr Brightwell's studio, which had a strange musty smell, was full of these bones and stark mounted skeletons which had bleak personalities of their own. Larger skeletons, including that of an elephant which had been brought from London Zoo bone by bone, resided in the loft. The local postman nearly fainted one morning when he reached into his bag and took hold of a Chimpanzee's hand which had become unwrapped. Mr Brightwell, tall, gaunt and bespectacled with long grey hair which fell over his eyes, was an important authority on wildlife anatomy but locally was regarded as an eccentric. Out of gastronomic curiosity he had cooked and sampled the meat of many of his subjects. He had a great sense of humour and, whilst some preferred to stay clear of him, I greatly enjoyed his company. In fact I often worked with him and at one time spent weeks doing some of the routine drawings for an animal cartoon film he was preparing. This association, which lasted many years, may well have been responsible for kindling my later interest in drawing wild birds. Certainly it resulted in my preoccupation with anatomical correctness.

It was not until the mid–1950s that my interest in birds really began to develop. I had been in the army and later a ship's purser with the Orient Line, but had now left the sea and was working with the Cunard Steamship Company in London. Weekend visits to the woods and fields took on a new importance. Now I was looking more objectively at birds and animals. Many of my lunch breaks were spent looking at the ducks in St James's Park which were a tranquil escape from the rigours of a busy London office. At this point my interest in birdlife and artistic ability came together. At first the two were hardly compatible. The rigidity of mechanical things and the softness of feathers bore no relationship. Nor did the exaggerations of caricature and cartoon favour the precise nature of bird character. My early attempts were less than satisfactory but I was determined to develop a technique which would enable me to paint these elusive creatures. I soon began to notice that the subject I had chosen did not command the same respect as my planes, caricatures and cartoons of the past. At the time birds were still regarded as the province of the country clergy and eccentrics. It was not the kind of thing a young man was expected to interest himself in. I might just as well have been learning to knit!

When I married my wife Daphne in 1960 I had no further concern about the opinions of others because she liked my work and that was all that mattered. Hitherto I had of course been encouraged by my parents even to the extent that they had a small studio built on to the house for me but, like most people, they were doubtful about the possibilities of making a career out of art and especially bird paintings. They were right to have such reservations because I could not find anybody, in Britain at any rate, who had made a living purely from bird art. I found that the small handful of artists who had specialised in bird subjects before me either had private means or other occupations which allowed plenty of time for painting. I could find no precedent which suited my needs. In fact I was advised to become a clergyman, teacher or perhaps go into forestry. I could not see myself in any of these roles and so continued to work in London for four more years, painting in the evenings and at weekends.

My first exhibitions were at the Tryon Gallery in London. In opening the first specialist wildlife gallery ever, the Hon. Aylmer Tryon was chided by his elder brother, Lord Tryon, for risking such a venture. The dawning of public interest in wildlife had not yet arrived. In the event, his foresight was sound and Aylmer Tryon was right to open his magnificent Dover Street Gallery. In 1964 the time was also ripe for me to give up my London office job and concentrate full time on bird painting. In that year I had to prepare an exhibition at the Tryon Gallery, another at the National Collection of Fine Arts of the Smithsonian Institution in Washington D.C. and, as if that were not enough, I also had to prepare thirty-six plates for a book.

Overseas exhibitions are expensive to finance, especially coming in one's first year as a professional painter. The Washington exhibition was to be a prestigious début in the New World, especially as it was the first ever to be given to a living artist at the National Collection. My father-in-law, Frank Parker, had implicit faith in my ability and provided generous financial support to see me through this first year. He had once been a professional artist himself and therefore understood the kind of problems this entails.

It is one thing to enjoy seeing a bird in the wild but quite another to convert the experience into a picture. A good artist is capable of doing a competent picture of almost any subject, simply by translating visual shapes, textures and colours onto paper. However, to develop an understanding of one's subject, which takes years of devoted study, is to add a spark of life that might otherwise be missed. Although I have been painting birds for twenty-five years and have come to see them as individual characters, I am not aware of any real contact with them such as the relationship one has with a cat or dog. Birds are shy, elusive and independent. Their territories overlap our own and, for the most part our presence is resented. Herein lies their romance and mystery for when we do have a close encounter, such as a Blackbird who once insisted on joining us at the breakfast table each morning, we are amazed and delighted. We are charmed by Robins who will accept titbits from the hand, simply because we are witnessing a momentary relaxation of the laws of the wild.

In Victorian times it was fashionable to keep songbirds in cages to delight the eye and ear but they were pathetic in their imprisonment and if they did sing, it was from instinct rather than contentment. And so it is whilst flying free in their natural habitat that we can most enjoy the presence of birds.

As birds are detailed creatures I like to do my paintings in great detail also. I try to convey my enjoyment of the feel of feathers and to capture the subtle presence each bird has. Of course they will not sit for you and it is impossible to paint from them direct, so I have had to develop a quick sketch technique to record as much information as possible in a few seconds. I use other aids such as binoculars and camera. It is not satisfactory to paint from photographs but they can often provide useful information which may have escaped the eye or the memory.

To my mind each bird has a decided character which needs to be conveyed. Actually birds have very little change in facial expression. With human beings the slightest nuance in facial movement is detectable and meaningful because we are programmed to receive and transmit in this way. With birds it is the movement of the body, deployment of feathers and song that convey the signals of communication.

For example: an angry Great Tit has its head drawn in and trembling wings half open; a cornered Kestrel lies on its back with claws advanced, spitting or hissing noisily; agitated Blackbirds 'chip' loudly with their tails jerking upward every second or so. The signals of romance are equally dramatic. I need therefore to know how the bird works: how its legs and feet are jointed; how far it can stretch or turn its neck and the way in which wings and tail are used. This kind of knowledge gives me the freedom to choose any pose I wish for my picture. The choice of bird arises from inspiration from the field or from simply being asked to paint a particular subject.

Having decided on a subject I begin to sketch the bird, very quickly at first, often with pencil held at arm's length. Satisfied that I have chosen a lifelike position I can begin to work more detail into the drawing, making careful measurements from study specimens or notes. The background is then suggested roughly. Some of my backgrounds are painted freehand and others, where there are elaborate plants for instance, are drawn out carefully in advance. At the same time I am bearing in mind composition and the balance of the picture as a whole. After that, all I have to do is fill in the colours! At this point I really become involved and the picture, which started off quite quickly, then goes through a series of processes from the rough to the painstaking and time-consuming finish. Most of the paintings in this book are fairly small and therefore did not take long to complete. The early ones, such as the Nuthatch on Silver Birch (page 103), took only a couple of days to paint. Such was the limit of my development at the time when I simply did not know how to apply more texture to my subjects. It may be interesting to compare the Nuthatch picture with the Greenfinches on page 35 which were painted eleven years later and took ten days or so to complete. By constantly learning more both about birds and about painting techniques much more interest is added to the subject.

Most of the work I do today is on a much larger format, measured in metres rather than centimetres. Such paintings can take anything from six weeks to three months depending on the complexity of composition and plumage. During this time I begin to feel an intimate relationship with my subject. I am, after all, painting it feather by feather, hair by hair and scale by scale. The brush seems to do marvellous things that I had not deliberately contrived. A spell has been cast and I am at one with my subject, ready to reveal its very being. The eye must seem capable of vision, the bill must feel like horn and the bird must have its inner soul intact beneath feathers that can all but be stroked. Even the body warmth must seem to radiate from the picture so that I can feel as if I am creating life itself. The successful conclusion of a painting brings an elation which is hard to describe.

My media is water colour. I once tried to paint in oils but found it too sticky for my purpose. I ended up with the stuff all over my hands, in my hair and over everything else too! Water colour, used with Chinese White, lends soft hues and textures that are ideal for birds' plumage. It is cleaner to use and dries quickly so that I can develop the work continuously instead of waiting for one phase to dry before going on to the next.

What makes an artist want to paint whilst others wish to be sportsmen, engineers, financiers or whatever? The uninhibited purity of children's painting, which I see in the work of my two young sons, gives us a lead. The best explanation is perhaps that art is a personal expression which records something we like or, that it creates and visualises a desirable experience from our imagination. I suppose that in my case, if I had all the birds I like in cages around the house, I would not feel the need to paint them for they would have lost their mystery.

I have a special affection for ducks and geese. A few years ago I was 'adopted' by a clutch of newly hatched wild Mallard ducklings. The parent birds had been shot (at least I had heard shooting in the vicinity hours before). The ducklings had formed a 'fur hat' in the garden of my parents-in-law and I had been summoned to come and see them. I bent down to look closely at the circle of yellow and brown fluff. It did not

move but a dozen pairs of tiny bright eyes opened and fixed me with a multiple gaze. As I stood up to walk away, the platoon aroused itself and followed me in line astern with tiny pink webbed feet going like pistons. Whichever direction I took they followed. As we were to move from Surrey to Sussex the very next day, I was not quite sure how to handle this unexpected responsibility. Utterly bemused by the predicament I was in, my parents-in-law agreed to look after the ducklings indoors whilst we moved to our eight-hundred-year-old house in the Sussex Weald.

During the first night of human care, some of the ducklings died. Although the room they were in was warm, it had not been quite warm enough and so the survivors were placed in the oven with the door ajar and the heat turned to low. It was an improvised incubator and a good idea because the diminutive creatures thrived and fed well on chopped, boiled egg-white. Eventually the young birds were delivered to us with the first signs of feathers beginning to appear. They became constant companions to my son Ashley, then three years old. We had no pond so they happily piled into a plastic washing up bowl. Later they swam in an old bath tub I had let into the ground for them. I also made a pen to keep them in at night because we had seen Foxes in the area. Each day I dug over a piece of ground for them and they crowded round the spade to pick out worms and insects, but their main diet consisted of duck pellets purchased from a local store. Every few days it was necessary for me to sally forth with buckets to find fresh Duckweed from wayside ponds. They awaited this with eager anticipation and became very excited when it was put into the 'bath' water. They clamoured for food when hungry and, if I was indoors, they gathered to beat a noisy tattoo on the glass door of the house.

In full plumage my ducks were magnificently sleek and glossy but they had not yet learned to fly. After all, nobody had shown them how. Endless wing-flapping went on but no attempts were made to get airborne. I decided I would need to give them some elementary instruction so, one evening, I set them at the end of a suitable runway and ran along it with my arms flapping. With necks craned they looked at me in utter amazement. A pet rabbit belonging to Ashley, which was then only a few weeks old, watched the lesson with raised ears and twitching nose. After several attempts the ducks followed me along the 'runway' flapping wings with gusto. So too did the little white rabbit, who had been brought up with the ducks, only he tried to flap his front paws. Neither I nor the rabbit managed to get airborne but the ducks did, and then I had to go round fields and neighbours' gardens to collect them. Their early flights were hilarious for they overshot the runway, landed untidily in bunches of feathers and generally demonstrated their lack of proper tuition. After several more months the flights, which tended to take place at sundown, became longer and more spectacular. Before take-off necks craned skywards, and then at a given signal, they took off vertically to make arrowhead formations high above. After a circuit or two they set course for some feeding ground on the Pevensey marshes a few miles away. Each morning, at sun up, they returned, usually in pairs, to become Ashley's pets again, sleeping lazily in the mid-summer sun or occasionally foraging about under the bushes.

One morning there was a duck missing. I was afraid it might have fallen prey to a hunter's gun or some other predator whilst out on the marshes. The others had returned at the usual time and whilst most of them settled down to rest, one took up station outside my glass door. It was unusual behaviour so I went out to investigate. In obvious agitation, the duck flew a few yards in one direction across the front garden. It came back and repeated the exercise until, out of curiosity, I followed. It finally led me quite a distance until, hidden under a bush, I found the missing duck. It had been injured, possibly by a power line, when coming in to land. It could not fly but did not seem to be too badly hurt. I penned it up with its companion for a few days until it was completely recovered.

Eventually all the ducks took off one evening never to return, collectively at any rate. Occasionally one of them would drop in for a quick visit and then fly off again. The call of the wild had overcome their domestic beginnings and we were sad to see them go. Nevertheless they were free to go their way and it had been an extraordinary insight into the domestic lives of wild duck. In treasuring this memory I regard all waterfowl with the utmost respect. Every duck painting I do is instilled with the personalities of these marvellous creatures of

whom I was privileged to have temporary guardianship.

Encounters with birds invariably provide interest and excitement. We are still learning about them and even from them. Sophisticated navigational aids have been developed relatively recently in Man's history, yet birds have had them since time began. What kind of micro-computer can bring Swallows unerringly back to last year's nests, the birds having travelled thousands of miles across continents in the meantime? Of course we know they use the stars but how do they work out the coordinating factors? What kind of emotion can cause a Canada Goose to die of heartbreak after its mate has been shot? What feeling prompts a bird to decorate its nest with small flowers in order to attract its mate? What kind of strength can enable some birds to fly the oceans? We cannot fail to be impressed by these breathtaking aspects of the world in which we live.

I accept the presence of wild birds with interest and pleasure, but although I am a romantic, I am not a sentimentalist. The lives of birds can often be harsh and frequently cruel. They observe strict natural laws and are fully extended in their role. Their life patterns are complex yet well ordered, and they have developed remarkable abilities that are sometimes beyond our comprehension. All their actions are geared to survival and procreation. Why are they so important to us? Put simply, the answer is that they, like ourselves, are part of nature's delicate balance. In the 1950s, because of the use of toxic chemicals in agricultural pesticides, Barn Owls practically disappeared from our islands, and are still scarce today. They, and other birds of prey, were literally poisoned by the thousand. The country has recently suffered huge infestations of rats and mice which would otherwise have been controlled. This means risk of disease for Man. Small birds keep insect pests at tolerable levels or take the seeds of many unwanted weeds. On balance we derive direct benefit from the presence of birds yet we offer considerable threats to their existence. Certainly there is a greater awareness in recent years, largely brought about by the work of the Royal Society for the Protection of Birds, the media, and dare I say, bird art, whilst large chemical manufacturers are doing a great deal of research to provide fertilisers that are harmless to Man and birds. However, oil pollution along the coast kills thousands of birds each year and erosion of habitat, which means any disturbance to the land, has an incalculable effect on bird populations. With the high cost of fossil fuels large acreages of woodland are being felled for logs. This would not be so damaging if we adopted the American methods of strip-cropping replacing trees in the resultant gaps. England simply cannot afford to lose more trees and hedgerows for not only does this affect the wildlife but also the weather. In my immediate area, where hundreds of acres of wood and copse have been ravaged, the winds often gust at over a hundred miles an hour. They are no longer tempered by the trees. In France many of the small birds which might otherwise have reached our shores are slaughtered either for the cooking pot or simply to show off hunting prowess. One hundred million are destroyed each year in this way. Birdwatchers and ramblers, with the best of intention, disturb breeding areas that might otherwise have remained secluded. As the most heavily populated country in the world with 800 people per square mile, as against 50 in America, perhaps we should realise we have a special responsibility.

One man who understands the responsibility we have as guardians of our natural heritage is Robert Dougall. Whilst I have said all I know how to say about birds with my brush, Robert Dougall paints his pictures in fresh, inspired words. His professional life has been devoted to the spoken and written word and so his thoughts and recollections are illustrated in a way which evokes the very essence of the countryside. There can hardly be a family in Britain which has not enjoyed Robert's reassuring presence in its living room during his years with the BBC. It is equally reassuring to journey through this book with him and enjoy the true romance of the countryside which he loves and understands.

Working with Robert Dougall in producing this book has been an enjoyable experience. Because we have both worked for long periods with large organisations in the metropolis perhaps we have a special appreciation of the countryside. In his attitude towards wild birds and animals I find a kindred spirit. He is dedicated to the conservation of our natural environment and, as former President, now Vice-President of the Royal Society for the Protection of Birds, he has played an important role in furthering the awareness and enjoyment of wild birds we have today. Perhaps this book, with its pictures and words in peaceful harmony, will offer yet more insight into a subject that can add an extra and very special dimension to our lives.

Basil Ede

Herstmonceux
Sussex 1980

The plates

Blue Tit

Lithest, gaudiest harlequin!
Prettiest tumbler ever seen,
Light of heart and light of limb,
What is now become of him?

William Wordsworth

This question is often asked when the Blue Tits, which have been welcome guests in gardens during the winter months, vanish as the breeding season approaches. Their charming antics around the hanging food-containers are then sadly missed. The answer of course is that they return to neighbouring woodland, where in most years the oaks provide a plentiful supply of caterpillars for their young.

Here, Basil Ede catches all the rapture and promise of early spring. The male, on the left, has fractionally more vivid colouring. How odd that such a sprightly, mischievous little creature should ever have been known as the Nun; and yet the blue cap and white, fillet-like cheeks must have earned him the name. Some of his many other local names are more readily understandable – Tom Tit and Billy Biter among them. The latter no doubt arose because of his aggressive behaviour on the nest, when any interference is met with hissing and biting.

Blue Tits rival Robins in the oddity of the places in which they choose to nest. Back in the days of highwaymen, there was even a report of a nest found in the jaws of a skeleton left hanging on a gibbet. At the same time, there is no bird that will more readily make use of a garden nesting-box. The great thing to remember is to site the box away from the full glare of the sun and, even more important, to make sure that a cat will not be able to reach it.

In the autumn Blue Tits sometimes behave rather strangely. When in Suffolk I often sit up late writing in the attic of my cottage. One night I was mystified by tiny tappings at the window. Then, through the leaded lights, I saw the pallid countenance of a Blue Tit. It may be that it was simply pecking at the fish-oil contained in the putty around the window panes. On the other hand, I know I shall long remember those 'tip-tap nothings of a tiny bird' in the silence of the night.

Hawfinch

The biggest of our finches is also the rarest, and how handsome he is on the sprig of wild rose. Basil Ede's bird is positively vibrant with life: see the lively glint in his cherry-red eye – it shows how wary he is. The Hawfinch is adept in the skilful use of cover and, I must admit, this is the best sight I have ever had of one. So let us have a good look at him while we may.

Sir Thomas Browne, the seventeenth-century Norfolk naturalist, described the bird as being: 'finely coloured and shaped like a Bunting. It is chiefly seen in summer, about cherrietime.' Apart from the splendid colouring, the plumage has a peculiarity: some of the primary feathers forming the white wing-bar are curved and squared at the ends, possibly to strengthen his flight. The main peculiarity, however, is of course that phenomenal bill, which extends the curve of the head making it look as business-like as a bullet.

All finches are seed-eaters, but the Hawfinch's speciality is cracking cherry stones to get at the soft kernel. Experiments have shown that it takes a pressure of up to forty kilograms to fracture the shell, so he really does pack a punch. He also has a great liking for peas and it may be as well for vegetable-growers that he remains rare. On the ground he hops awkwardly and looks strangely top-heavy with his massive head and short tail. A strong, silent character indeed.

Bullfinch

See the bright bullfinch now,
High on the apple bough.

Anthony Rye

There is no doubt of the Bullfinch's beauty: he looks so burly and proud. Here, Basil Ede shows the male to perfection in all his breeding finery; and what a thrill it is to glimpse that glossy black cap and near blood-red breast glowing in early spring sunshine. It always seems to me that he has something of the look of a dashing, if diminutive hawk.

And yet when he attacks the spring blossom leaving the grass beneath the fruit trees and ornamental garden shrubs littered with tender young buds, it is enough to try the patience of the most devoted bird-lover. It is some consolation to know that, as he mainly eats seeds in winter, he will not attack the fruit buds nearly so much in those years when the seeds of nettle, dock, bramble or ash are plentiful.

Although many Bullfinches have to be destroyed by fruit-growers each year, they continue to increase in numbers. This may partly be due to the extra food supply they derive from the bigger acreage of cereal crops.

When not despoiling the fruit, he is an admirable character, shy and retiring. Unlike most small birds, a pair will stay together all through the winter. A menace in the orchards he may be, but who can help rejoicing at that first glimpse of him in spring?

Chaffinch

The Chaffinch is a bird of mixed woodland and farmland, a lover of hedgerows, orchards and parks, but less at home in small town gardens and surburban plots. In my tiny Hampstead garden he is quite a rarity, whereas in Suffolk I scarcely give him a second glance. And yet, as one of our commonest birds, how fortunate that he should be such a merry character and above all, such a good-looker 'like to a sunbeam made of coloured wings'.

Strangely enough the Chaffinch has been comparatively little studied or celebrated. Perversely, it is often only when deprived of something that we fully value it. For Robert Browning, languishing abroad, it was the memory of the Chaffinch's simple song that made him ache for home:

> Oh, to be in England
> Now that April's there,
> And whoever wakes in England
> Sees, some morning unaware,
> That the lowest boughs and the brushwood sheaf
> Round the elm-tree bole are in tiny leaf,
> While the chaffinch sings on the orchard bough
> In England – now!

Yes, the Chaffinch deserves better of us than simply to be taken for granted.

How colourful and handsome is the male, seen here in his breeding plumage; the hen drably modest in comparison. During courtship the cock makes a great display of his breast which glows the colour of a mellow brick, and he turns proudly from side to side to show his white wing patches. He also raises a tiny crest. At this time he is full of fight, especially against males of his own kind. While staying in Suffolk in early spring, I remember one seemingly demented bird constantly attacking his image in the off-side wing-mirror of my car, parked outside the cottage. I had to throw a duster over the mirror or I believe he would have knocked himself out.

The Chaffinch's joyous, defiant song is first heard in February, although a few weeks may pass before he gives it the final flourish. It lasts only two or three seconds and with its persistent reiteration is one of the most easily distinguished of all our bird songs. No wonder Browning missed it so much.

Little Owl

Compared with his relations, the Little Owl undoubtedly has a fiercer mien, but then he is an alien — and a controversial one at that.

His British career started back in the nineteenth century when various attempts were made to establish Little Owls here from Continental populations. By the turn of the century these had been successful. They spread rapidly and soon every gamekeeper's hand was against them. Little Owls were accused not only of taking game chicks but of snatching songsters as well. Before long the press also joined in the chorus of condemnation.

As a result, in 1936 the British Trust for Ornithology set up an inquiry with Miss Alice Hibbert-Ware in charge. After analysing the contents of 2,500 pellets from thirty-four countries, the findings showed conclusively that throughout the years the staple food of these much maligned birds consisted of rodents, shrews, earthworms, beetles and insect larvae. Dr Collinge, their economic ornithologist, summed up: 'The bulk of the food is of such a nature that the Little Owl must be regarded as of great value to the agriculturist. As a factor in the destruction of injurious insects and voles and mice, it is a most valuable ally.'

Exoneration seemed total; yet, in spite of all this, a slight doubt continues to linger about the Little Owl as compared with the other members of his family. He frequently hunts in the daytime over open country where there is little cover, and so has the opportunity of killing small birds if so minded. And in some parts of the country he is still considered a foreigner.

Anthony Rye encapsulates the loneliness of this strange bird in a slightly macabre poem where he envisages the Little Owl haunting a remote burial mound in search of fellowship. But the dead know him not:

> For though the Little Owl,
> Squatting on stump or stone,
> Weeps like an earth-bound soul,
> He is to them unknown.
> He does not cry to come
> To other gale-blown trees,
> More calcined hills, of Spain
> That was of old his home;
> But, spectre-flitting, sole,
> He calls, and claims in vain
> The fellowship of these:
> Ghosts, who know not his name!

BASIL EDE

Cuckoo

'Cuckoo! Cuckoo!' the first we've heard!
'Cuckoo! Cuckoo!' God bless the bird!

T.E. Brown

There he is, the messenger of spring, shouting the good news from an apple tree bough. One of nature's layabouts he is sly, merciless and utterly devoid of morals. Unsavoury character he may be, yet when those two marvellous hollow-sounding notes ring out one cannot help but listen and rejoice.

The Cuckoo's breeding behaviour is both remarkable and unique. The male is the first to arrive in this country and soon sets up a territory. The female is silent on arrival, but when she is ready for what Chaucer charmingly described as 'a spring observance', she gives her strange, water-bubbling cry. She then sets about her devilish work. Having carefully selected the nest of another bird, she takes one of the eggs in her bill, lays one of her own and flies away with the stolen egg. It takes her precisely eleven seconds.

When the Cuckoo's egg hatches, the blind, naked infant proceeds with fiendish and instinctive ingenuity to heave the other eggs, or nestlings, out of the nest. From that time on, the wretched foster parents lavish care on their monstrous charge and have to work flat out to provide the intruder with food. After three weeks it will have increased its weight fifty times, be fully feathered and ready to leave.

I had the good fortune to see such a youngster in the garden of our Suffolk cottage in early September, 1975. It was perched awkwardly on the low branch of a wych-elm, every now and again fluttering clumsily to the ground in search of food. But it must have been smarter than it looked, because it soon caught an enormous earthworm and hoisted it in laboriously.

I knew that the carefree adult Cuckoos had already left for their winter quarters. It was incredible to think that in only a week or so this inept, gawky youngster would be winging its way for the very first time, all those thousands of miles over unknown land and sea to Africa.

Blackcap

The Blackcap is a singing bird,
A nightingale in melody.

John Clare

What could be more joyous and sprightly than this painting of
Blackcaps in early spring? How perfectly the clean, bright yellow of
the forsythia sets off their sober plumage. Unfortunately, the bird is
seldom seen as clearly as this and must be the shyest of the common
warblers, usually choosing dense cover from which to pour forth its
rich and varied song.

John Clare, the most observant of all our nature poets, saw his
Blackcap in a tangled hedgerow:

Where ivy flapping to the breeze
Bear ring-marked berries black as jet.

He goes on to wonder whether the bird takes berries for food. In fact,
unlike other warblers, the Blackcap does feed on berries when he first
arrives in spring, although insects taken on the wing are his principal
food. Later in the year he also likes soft fruit.

In some parts, the Blackcap is known as the Northern Nightingale,
but in my opinion, he is not to be compared with that bird, the greatest
of all our singers. Gilbert White of Selborne described the song as 'a
full, deep, sweet, loud, wild pipe' and rated it as superior to those of all
the other warblers, with the exception of the Nightingale. One thing
beyond all question is that the first sound of those rich, melodious
notes on a bright April day means that spring is truly here.

Goldfinch

I love to see the little Goldfinch pluck
The groundsel's feathered seed;
And then in bower of apple-blossoms perch'd
Trim his gay suit and pay us with a song.
I would not hold him prisoner for the world.

James Hurdis

Today everyone would agree with Hurdis who wrote this song of praise about 200 years ago. However, the Goldfinch's freedom has not always been valued so highly. By the turn of the century, the beauty of these birds had made them extremely rare; in 1860, 132,000 Goldfinches were trapped at Worthing alone, to end up in cages.

Fortunately, the Society for the Protection of Birds was formed in 1891, and one of its first successes was the promotion of legislation to stop the liming and trapping of songbirds. The Goldfinch has since made a great recovery.

How delighted Thomas Hardy would be for one, to see a little family flock or 'charm' moving about the countryside, the birds falling like sudden golden rain on the thistle heads they love. It is perhaps worth remembering that in Hardy's day it was all very different. He once wrote:

Within a churchyard, on a recent grave,
I saw a little cage
That jailed a goldfinch. All was silence save
Its hops from stage to stage.

Pied Flycatcher

Tree, tree, tree, once more I come to thee.

Anon.

That is how a countryman set the lively little song of the Pied Flycatcher to words. The song, an often repeated sequence of notes, has a friendly sound like a soft whistle. Living in the south and east, I feel distinctly deprived that these dapper, attractive birds should mostly confine their breeding activities to Wales and its bordering counties. They are also found, although to a lesser extent, in parts of Yorkshire and southern Scotland and in some other favoured spots in the Pennines, the Highlands and the west of England.

Almost the only chance I ever have of seeing one is in the autumn, when small parties move along the east coast *en route* to their winter quarters in Africa. Sometimes in September, when the winds come from the east or south-east, there can be quite large falls of birds migrating from Scandinavia.

In Basil Ede's painting, the cock, seen on the right, is in his smart spring plumage. After the autumn moult, the sexes are more difficult to distinguish, as the black of the male is replaced by dark brown and the white parts are less conspicuous.

Pied Flycatchers, like their more numerous relations the Spotted, have short legs as they seldom walk. Their movements are brisker, however, and the tail is constantly flirted or swayed up and down. They are not such brilliant performers as the Spotted at taking insects on the wing and will often drop to the ground to take a caterpillar or beetle, but rarely return to the same perch.

The favourite breeding places are in woods of oak or ash beside fast-flowing streams. Like the tits and redstarts, they usually place their nest in a hole in the decaying branch of a tree. Competition for these sites is keen and so the Royal Society for the Protection of Birds has provided numerous nesting-boxes for the birds on its Welsh reserves. The Pieds took to this accommodation enthusiastically and, in years when caterpillars are plentiful on the oaks, as many as 900 fledglings have been reared in the boxes on the Gwenffrwd reserve alone.

BASIL EDE 1977

Willow Warbler and Chiffchaff

Chiff-chaff, chiff-chaff,
After labour, rest I have...

Anthony Rye

These two vibrant notes are among the most welcome sounds of spring. The Chiffchaff is the first of all our warblers to dare the chilly winds of early March after wintering in North Africa. The sound of his song has been likened to two tiny taps of a hammer on an anvil. Monotonous it may be, but there is no doubt of the ringing vigour; he sings from his heart.

His close relation, the Willow Warbler or Willow Wren, has an even longer journey from the very centre of Africa and arrives about a fortnight later. As you can see the birds are difficult to tell apart. Here the Willow Warbler is the topmost bird, slightly brighter in colour and its legs a paler brown.

How strange that two birds so alike in appearance should have such totally distinctive songs! There is an ethereal quality about the sound of the Willow Warbler's song – a cadence of tiny, silvery bells, starting with round, pure notes and falling away to the gentlest murmur. Many writers have tried to capture its essence, among them the great nineteenth-century naturalist Richard Jefferies: 'So gentle, so low, so tender a song the Willow-Wren sang that it could scarce be known as the voice of a bird, but was like that of some more delicate creature with the heart of a woman.'

How lucky we are that this melodious, graceful little bird should be found almost everywhere in the British Isles in summer and especially where there is birch, oak, hazel or hawthorn. The name is misleading, as the bird is not particularly partial to willows and he is certainly not a wren. All one can say is that of his two principal names Willow Warbler is the least inaccurate.

Similarly, it is a misnomer to call the Chiffchaff a warbler; some might say he is more of a chirper. But there is no doubt whatever of the eager welcome he gets each March, when those two joyous notes hammer out in the woods and we greet again the harbinger of spring.

Greenfinch

Thou, Linnet! in thy green array,
Presiding spirit here to-day...

William Wordsworth

Wordsworth uses here the older, regional name for the Greenfinch of Green Linnet and Basil Ede's painting is a perfect accompaniment to his words:

Beneath these fruit-tree boughs that shed
Their snow-white blossoms on my head
With brightest sunshine round me spread
Of spring's unclouded weather...

In the splendour of his spring plumage the cock, seen here on the left, is indeed a lovely bird; the hen's colouring is a little duller. Of all our eighteen finches, after the Chaffinch, the Greenfinch is one of the commonest, and certainly one of the least shy.

This pair, so spry among the blossom, may well have survived many cruel winter days thanks only to the peanuts on countless suburban bird-tables. Fortunately they are well built for survival: in our Hampstead garden one of the regulars last winter was a real toughie who drove off all-comers, even House Sparrows and starlings.

There is a softer side to the bird's nature which is revealed in courtship. In early March I have often watched a cock take off from a topmost spray in hovering, circular flight, and listened to his tiny, bell-like notes before he dived back into the cover of leaves again. Then, on the ground, with trailing wings and head forced back, he displays to the hen the full glory of his gilded breast. She, meantime, crouches down shivering her wings and begs for food like a fledgling.

Graciously he thrusts his bill into hers and regurgitates a whitish substance from his crop. He will also feed her later when she is sitting on her eggs and when the young hatch out both parents feed them, again by regurgitation.

At the end of the breeding season, in bounding flight, some Greenfinches will join the flocks of Chaffinches and buntings feeding on wasteland and stubble.

When it comes to identifying a Greenfinch by his song, there is one unmistakable, drowsy, drawn-out note he delights in on the hottest days of summer, when all but he are still. It may be wishful thinking, but it sounds almost as if it could be 'bree-ee-eeze'.

Mallard

From troubles of the world
I turn to ducks...

F.W. Harvey

The sudden sight of a newly hatched brood of Mallard ducklings in the reeds will surely melt any heart. There may be any number from seven to sixteen or even more, so Basil Ede's downy, bright-eyed brood is quite a modest one. The mother duck is a devoted parent, unlike her raffish mate, and at this moment, may well be beating the water with her wings and quacking like mad to draw attention to herself and away from her threatened young.

It is she alone who will have sat patiently on the eggs for all of the twenty-eight days incubation, stealing away only for food, and that usually under cover of darkness. She will then carefully cover the eggs with soft, dark down plucked from her breast. When at last the young hatch, they leave the nest at once, running and scampering over the water like tiny clockwork toys. It is then that the mother duck's real worries begin.

Below water, a pike may be lurking to seize them; on land, there are rats, weasels, stoats and foxes, while, from the air, sudden death may come from scavenging crows, gulls or even a lumbering Heron. Sometimes, the nest is as much as nine metres up in the fork of a tree. The tiny bundles then have to launch themselves into the unknown, floating to the ground like thistledown. What a marvel is a duck!

Canada Goose

There swims no Goose so grey but soon or late
She finds some honest Gander for a mate.

Alexander Pope

This engaging trio of Canada Goslings has been hatched only a few days. They look snug and secure on their downy bed among the dry grasses and dead leaves of the nest, yet they may well be the survivors of a brood of six or seven and will already have had countless adventures and hair-breadth escapes. From the moment of smashing out of the egg, covered in golden down and with eyes open, they have been learning: first to follow their dam, albeit waveringly, down to the water on webbed, baby feet; and then to paddle after her, stroke after tiny stroke. At least they will not have had to worry about food; their bodies contain enough yolk from the eggs to sustain them for some time. It will be six weeks before the adventure of flight.

The Canada Goose is common in North America where there are a dozen or more different races, and it was introduced into Britain in the seventeenth century to grace our ornamental lakes. A census taken in 1953 put its numbers here at between two and four thousand, but I am sure there are now many more of these big handsome geese with their glossy black necks and distinctive white patches on cheek and chin.

Tree Sparrow
and
House Sparrow

In busy mart and crowded street
There the smoke-brown sparrow sits.

Eliza Cooke

In town the House Sparrow is about as nondescript as it is possible for a bird to be; but see him in clean, country air in spring and he's quite a beau. Here he is in the middle with his rather drab little hen below him. The topmost bird is his close relation, the much rarer Tree Sparrow, with a chestnut crown and black spot on each cheek.

Sparrows are related to the African weaver-birds and must have spread north across Europe, when primitive Man first began to cultivate crops. As a seed-eater, the House Sparrow quickly moved in on Man's dwellings, where lodging and easy pickings were always to be found. This also relieved him of the need to make dangerous migration flights – in fact Spuggie, as he is sometimes known in the north, had got it made. And he's been making use of Man ever since.

'Yea, the Sparrow hath found her an house' wrote the psalmist, and a pair has nested in a clump of clematis just outside the bedroom window of my Hampstead cottage for several years now. My wife and I have simply had to learn to put up with their incessant cheepings.

Perhaps Spuggie's biggest break came in the middle of the last century, when the citizens of Philadelphia were plagued with span-worms which were destroying the shade trees and dropping down people's necks. England came to the rescue; a thousand sparrows were transported and released in the heart of the city to a warm welcome, but not for long. In no time the complaints were bitter. The birds were multiplying exceedingly, eating quantities of corn, and usurping the American songbirds, whom they allowed to remain in their haunts 'only on sufferance and after much battle'. The House Sparrow had taken a leaf out of W.C. Fields's book: 'never give a sucker an even break'.

Since then, with Man's help, Spuggie has spread to Mexico, Australia, New Zealand, South America, Hawaii, Mauritius and the Falkland Islands. Success indeed.

Tree Sparrow ~ *Passer montanus* (top)
House Sparrow ~ *Passer domesticus* (male & female)
© BASIL EDE 1979

Lapwing

Whizz goes the pewit o'er the ploughman's team
With many a whew and whirl and sudden scream. . .

John Clare

I must own to a special affection for the Lapwing, which has always been a favourite with the countryman, hence the variety of its local names. I always thought that the name Lapwing referred to its tumbling flight, but scholars say it derives from the Old English word 'hleapwince' which means 'a leap with a waver in it', and there is certainly no more perfect description of the Lapwing's flight than that. The bird is also widely known as the Peewit or Green Plover. Among some of the rather charming local names are: Flopwing, Lipwingle and, referring to his cry, Wallopie Weep.

When in Suffolk and living close to the reedbeds I am privileged to see the courtship flights in early spring. Once the reeds are harvested there remains a low, stubbly growth on the black mud, through which the new green shoots soon begin to show. It is perfect nesting ground for Lapwings.

The male flies to and fro on rounded wings in a crazy, twisting flight, then swoops up almost vertically and flips over in a back somersault, as he throws to the wind his reedy defiant cries. His mate on the ground appears to take no notice, but suddenly she will join him in an ecstatic, tumbling love-dance in the air.

The eggs were once considered a delicacy. The then infant Royal Society for the Protection of Birds carried out a campaign against eating them, however, and in 1930, received support from no less a person than King George V. *Bird Notes and News* declared: 'The action of HM the King in declining to accept the early Plovers' eggs sent to him this spring has been received with general acclaim.' Happily the bird is now fully protected.

If ever I had to leave Britain for good, I should like to take with me a recording of the Lapwing's wild cry. Then, I should be able to say with Anthony Rye:

I see the evening skies;
I smell the fresh-turned loam;
The same skies, the self-same fields
the very clods of home!

Kingfisher

It was the Rainbow gave thee birth
And left thee all her lovely hues.

W.H. Davies

The Kingfisher is certainly Britain's most brilliantly coloured bird, although often all we see of him is a flash of tropical blue, as he skims, straight as an arrow, over the water. It must have been beginner's luck with me, because the very first time I entered a hide on the RSPB Minsmere Reserve, I gently raised the shutters, and – there was a Kingfisher – hovering, almost vertically, a metre or so in front of me. That was thirty years ago but I remember it as though it were yesterday. Never have I seen anything look so vibrantly alive.

Since then, I have seen the halcyons many times over the dykes between my Suffolk cottage and the sea. I have even seen an occasional one and heard its high-pitched whistle over the Hampstead Heath ponds, four-and-a-ha miles from the centre of London.

Basil Ede's bird has characteristically chosen a dead branch so that he may have a clear view of the water below. As he is cautious he will not fish until he is sure the coast is clear. Then he will sit upright, tail downwards, head turned intently to scan the water; a sudden drop, a tiny splash, and he will emerge with his struggling catch. This he usually holds crossways in his bill as he flies back to his perch or to a nearby stone, beats it into stillness, and swallows it head first.

The nest is a hole tunnelled out of the bank of a stream or by the side of a canal or gravel pit. A mill-pond is also a favoured place. Sometimes a hole made by a Sand Martin or water vole is taken over. The tunnel usually stretches upwards for about a metre and at the end is the nesting chamber. If you should ever find one however, be warned! There is usually a noisome stench of decaying fish and the tunnel itself becomes a sewer of green slime. What a dismal start in life for the Kingfisher nestlings! Pink, naked and helpless, there they lie: a pile of prickly fish bones for a cradle. And how astonishing that from such a fishy stew there should emerge this bird of electrifying and startling beauty.

Puffin

A pair of Puffins are contemplating the serious business of breeding: melancholy clowns with gaudy noses and eyes permanently surprised. They look so bizarre that their creator must surely have a delicious sense of fun. Their noses not only appear false, but in a sense, they are. The brilliantly coloured casings peel off after the breeding season and their bills are then a dull yellow and smaller. However comical they may appear, the Puffin is nevertheless, a highly efficient member of the little auk family.

After seven months wintering at sea, the Sea Parrots as they are sometimes known, return in March or April for the most important occasion of their year. At first they form 'rafts' on the sea close to the breeding site, but after a few weeks, most of them are ashore and busying themselves tunnelling into the soft, sloping turf on the cliff-tops with their pickaxe-like bills. Sometimes they seem suddenly to materialise on the cliff-tops, standing erect as though arriving for a high-level, dress-suited conference with their glistening white fronts, glossy black backs and those incredible bills. At the end of the metre-long tunnel, there soon lies a single, white egg. Both parents, although mainly the female, then set about the task of incubation which lasts about six weeks.

At the end of this time, the parents can be seen flying back to the colony with as many as six sand-eels or little fish, held crosswise in their bills. No one quite knows how they manage to retain the first fish caught, while opening their bills to snap up others. Ahead of the chick are six weeks of intensive feeding; then, it is suddenly deserted. For eight days it lives on its accumulated fat, until finally, hunger forces it up to the surface. On evenings in late July, the abandoned youngsters can be seen standing by the nest entrances looking distinctly down-in-the-mouth. Eventually they pluck up courage, waddle to the cliff-edge under cover of darkness, and – over they go – fluttering down to a 'life on the ocean wave'.

It is sad to think that the Puffin is now one of our most threatened seabirds; but it is good to know that the Royal Society for the Protection of Birds have now acquired several cliff-top sites as breeding reserves.

Puffin - Fratercula arctica
© BASIL EDE 1979

Jay

The Jay, the Pie, and even the boding Owl
That hails the rising moon, have charms for me.

William Cowper

The two fledglings in Basil Ede's painting are nearly three weeks old and ready to leave the small compact nest of twigs that they have almost outgrown. Their plumage is already unmistakably that of the gaudy Jay. The only differences are that the young have a brown iris to the eye, whereas that of the adults is a light blue, and their bills are much paler.

These are the two survivors of a clutch of anything from three to six eggs: it is usual for a number to disappear mysteriously at hatching time. Some think the hen bird is the one responsible, but the losses might equally well be caused by Carrion Crows, Magpies or owls.

Our two fledglings will soon be launching themselves awkwardly from the nest and whenever possible hopping and fluttering upwards. They seem to know that greater dangers lurk for them on the ground. Even when they are able to fly quite strongly, their parents will continue to feed them, and family parties sometimes remain together until the following spring.

Birds of the oak woods, they have in recent times taken to town life and now breed boldly in city parks. Many a Londoner has, I am sure, been first aroused to the wonders of birdlife by glimpsing the exotic beauty of a Jay.

Barn Owl

The Owl that, watching in the barn,
Sees the mouse creeping in the corn...

Samuel Butler

The heart-shaped face, the questing eyes, the silvery-white plumage dusted with gold all help to make the young white owl one of the most irresistible creatures in nature.

It is all the sadder that its numbers are declining especially in eastern England, for this bird is above all the farmer's friend. A pair of Barn Owls feeding young have been known to bring a mouse to their nest every ten or fifteen minutes.

Unfortunately, with the farmer's dedication to efficiency and tidiness, the countryside today has fewer old barns, ruined buildings and hollow trees in which the birds like to nest. A wise farmer will leave holes in the gable ends of his barns and put up a nest-box on the end gable wall; a tea-chest with a landing platform will be sufficient.

Recently, in Suffolk, my wife and I while driving spotted a young white owl hunched by the roadside. He had evidently been stunned after collision with a car. We stopped, wrapped him gently in a rug, and took him to a lady with an aviary and access to a supply of mice. Fortunately, his recovery was rapid and four days later he was released near the place where we had found him – flying away strongly on soft, silent wings.

Mute Swan

Oft have you seen a swan superbly frowning,
And with proud breast his own white shadow crowning...

John Keats

There are few more serene and majestic sights than that of a male Swan, the cob, escorting his young on a reedy mere. We know it is the male because of the large black knob at the base of his bright orange bill. In the female this distinguishing feature is less prominent.

Back on the nest in the reeds, she will still be sitting closely on the remaining four or five eggs. After five weeks they hatch at daily intervals. Each newly arrived cygnet remains on the nest for a day or two and then each, in turn, will be taken for a swim by the cob, until the whole family is able to take to the water in convoy.

There is a family each year on the dykes between my Suffolk cottage and the sea. To come upon them suddenly in that solitude always gives me a sense of privilege: it is, for a moment, as though nothing has changed since the world began. Wordsworth found perfect words for the scene and noted how the young are often carried on the female's back. Heading the stately procession comes the proud male:

He swells his lifted chest, and backward flings
His bridling neck between his tow'ring wings;
Stately, and burning in his pride, divides,
And glorying looks around, the silent tides:
On as he floats, the silver'd waters glow,
Proud of the varying arch and moveless form of snow.
While tender Cares and mild domestic Loves
With furtive watch pursue her as she moves,
The female with a meeker charm succeeds,
And her brown little ones around her leads,
Nibbling the water lilies as they pass,
Or playing wanton with the floating grass:
She in a mother's care, her beauty's pride
Forgets, unweary'd watching every side,
She calls them near, and with affections sweet
Alternately relieves their weary feet;
Alternately they mount her back, and rest,
Close by her mantling wings' embraces prest.

Mute Swan ~ Cygnus olor
© BASIL EDE 1979

Pied Wagtail

Little Trotty Wagtail he went in the rain,
And tittering, tottering sideways he ne'er got straight again.

John Clare

The elegant Pied or Water Wagtail must be the nearest approach to perpetual motion in the bird world. In spite of its name, it does not need to be by water and farmland suits it very well, although nowadays chemical sprays have destroyed much of the insect life on which it feeds.

On the small lawn behind my Suffolk cottage it is a delight to watch the Pied Wagtail's balletic display of dipping erratic flight. It seems to act on the impulse of the moment – running, standing, tail-wagging – interspersing this with sudden vertical leaps into the air.

When I worked at the Television Centre in London I was surprised to find that one of the regulars at the roof-garden restaurant was a rather grimy Pied Wagtail. I then received a letter from a viewer who remembered the district before the building was erected, and asked if the Pied Wagtail was still there. What tenacity the bird must have shown to survive all the upheaval – Pied Wagtails must be tougher than they look.

F.O. Morris in *A History of British Birds* recounts how earlier this century a Pied Wagtail built her nest on the framework underneath a railway carriage. The train travelled forty miles on a local run each day; meantime, the male used to await anxiously the return of the hen and his four little ones. Perpetual motion indeed!

Grey Wagtail

If the Pied Wagtail is elegant and constantly active, then the Grey is the ballerina *par excellence* of the family, or rather the naiad, because she is above all a water-sprite. The bird loves running water, the more turbulent the better, and is often seen in company with the Dipper, near waterfalls and rapids in the north. Here, Basil Ede shows a male in full breeding plumage, poised on a vantage point beside a fast-running stream. At any moment he may dart upwards to snatch a passing fly or flicker across the water from stone to stone, head and neck dipping forward, his ten-centimetre-long tail perpetually wagging up and down.

This is the summer scene but, come the winter, and the Grey is forced to move south. That is when I have often heard people say they have seen a Yellow Wagtail. The mistake is understandable, as even in the winter months the bird has a distinctly lemon-coloured breast. There need be no confusion, however, as the Yellow is only a summer visitor to Britain and leaves in the autumn for warmer climes south of the Sahara.

For me, it is always a special delight to come across a Grey in mid-winter. A couple of years ago, I even saw one elegantly bobbing its tail by a partly frozen pond on London's northern heights at Hampstead – in the rush hour.

Great Crested Grebe

Basil Ede's life-like portrayal of the Great Crested Grebe shows us one of Britain's most elegant and richly adorned waterbirds, which in itself, symbolises a triumph for the conservation movement.

The ancestry of the Great Crested Grebe goes back millions of years, as can be sensed from its primitive reptilian grace, and yet by the middle of the last century it had been brought to the edge of extinction. In 1860 a count throughout the whole country totalled only forty-two pairs. The bird had been a sacrifice to the millinery trade, nineteenth-century fashion dictating that women's hats should be vast winged creations, mausoleums of dead birds.

Fortunately, the Society for the Protection of Birds had been formed by a small group of ladies in 1891, and thirteen years later it received the Royal Charter. Legislation to control the plumage trade followed in 1921 and represented one of the infant RSPBs major victories.

Since then, assisted by new breeding places such as flooded gravel pits, the Great Crested Grebes have made a splendid recovery and can now be numbered in thousands. Surprisingly, one of the best places to observe Loons, to give the old Norfolk name, is on the ponds of London's Hampstead Heath. They have spread there from the network of reservoirs that ring the capital; in most years at least four pairs breed and it is a great privilege to watch their complex courtship display. The most impressive part is the so-called 'penguin dance'. Each bird dives to collect a billful of weed; they then dash towards each other over the water, suddenly rising up to full height, breast to breast, their heads swaying and the weed hanging down between them.

The nest is an untidy floating platform built of sticks and pieces of weed and, is usually close to the bank. The zebra-striped chicks can swim almost from hatching and spend much of the first few weeks snuggled down on their parents' backs. I know of few more fascinating sights in nature than watching the intimate family life of these strange and ancient birds.

EDE © 1976

Hoopoe

Whoop on, whoop on, thou canst not wing
Too fast or far, thou well-named thing...

George Darley

Well-named this exotic bird certainly is. I first heard its flute-like, hollow-sounding calls in a lightly wooded grove near the Pont du Gard in Provence. Then, close by the monumental stone aqueduct built by the Romans, a bird with broad rounded wings, like some outsize black-and-white moth, flew in slow dipping flight across the clearing.

Since then I have watched Hoopoes many a time in Africa and the Mediterranean countries and always with a feeling of delight: they are birds of exceptional charm and grace. On alighting the crest is momentarily flaunted like an American Indian's head-dress; then instantly lowered. As the bird diligently searches the ground for insects its bizarre beauty can be fully seen. There is something of the desert in the tawny-pink of its plumage, off-set by the black-and-white markings of the crest, wings and tail. Once, I watched fascinated as a Hoopoe tossed some choice morsel in the air, catching it elegantly on the way down.

The bird has an ancient lineage and has been much celebrated in painting and legend: the best known concerns the origin of its crest which dates back to the time of King Solomon. Once, when the King was journeying in the desert he was overcome by the heat of the sun. The Hoopoes took pity on him and shielded him with their wings. In gratitude, Solomon asked what he could give them in return. Misguidedly they asked for crowns of gold, as worn by Solomon himself. The birds departed joyfully, but were soon very nearly annihilated because of the gold they carried on their heads. Finally, the Hoopoes had to fly back to Solomon and tearfully beg for the crowns to be removed. This Solomon gladly did, obligingly replacing them with crests, which were even more beautiful.

In Britain the Hoopoe is on the extreme edge of its range and so seldom breeds here. The limiting factor is probably the scarcity of its favourite prey – grasshoppers, crickets and small lizards – in our colder climate.

Great Tit

Teacher, teacher, teacher,
That's your song.

Robin Ivy

'Tea-cher, Tea-cher' certainly gets close to the insistent, two-note sound of the Great Tit's cheerful hammering through the echoing woods: that most familiar sound of summer. An old name for the bird is Oxeye, which also resembles the often repeated low and high note; but it might equally well have stemmed from the look of the large white cheek-patch outlined in black.

Great and Blue Tits are by far the most common of our six members of the true titmouse family and their acrobatic antics make them star-turns at innumerable bird-tables in winter. The word titmouse, although descriptive, has nothing to do with small rodents but derives from an Old English word 'mase' meaning a small bird. In some ways it might have been better if Oxeye had stuck, as the other name sometimes gives rise to ribaldry and confusion.

Once, in the early days of the British Broadcasting Corporation, it was even said to have got the august Director General into trouble. A talk was being broadcast by an ornithologist who was listing foods certain birds liked. It so happened that a lady from the shires was entering her drawing-room from the garden as these four words issued from the loudspeaker within: 'Great Tits like coconuts'. That was enough for her! She rushed to the set, switched it off and sat down at her desk to pen a snorter to Sir John Reith accusing him of debasing womanhood, polluting the ether and heaven knows what else. Sir John, though shaken, replied urbanely: 'Dear Madam, if you had only continued listening you would also have heard that Robins like worms.'

Each winter in my Hampstead garden we have a merry company of Great, Blue and Coal Tits feeding on the peanuts and fat we provide for them in hanging containers. Wooden boxes with a wire mesh base, as sold by the Royal Society for the Protection of Birds, are by far the best for nuts as they can only be reached from underneath making it more difficult for competing sparrows.

Whitethroat

The happy whitethroat on the sweeing bough,
Swayed by the impulse of the gadding wind...

John Clare

Clare's words evoke for me the sight of the Whitethroat on a breezy day of early summer; and here Basil Ede's bird, swaying on a topmost stem among the tangled bindweed flowers, gives life to his words.

For years at the Suffolk cottage, my bedroom window looked straight out onto a high hedgerow formed by wild plum trees. It had grown unchecked and bordered a plot taken over by brambles and nettles. To watch the Whitethroat suddenly appear on an upswinging bough, pouring out a rapturous torrent of notes, which sometimes seemed to lift him into the air with joy, was one of my delights. All the more so because suddenly, for no apparent reason, he would duck down and vanish into the shelter of leaves.

Now the plot of land has a house on it, the hedge has been trimmed – so, no more Whitethroats. Fortunately there is still plenty of scrub left near me so the birds are not uncommon. All the same their numbers have nose-dived since 1968, when there were severe droughts in their wintering place, the southern Sahara. Millions died and their numbers have been slow to recover.

It would be a sad day indeed if the Nettle-creeper, to give one of its local names describing its movement through undergrowth, were no more to be seen in summer, swelling its white throat with song.

Nightingale

The music of the moon
Sleeps in the plain eggs of the nightingale...

Tennyson's magic words describe the greatest of all our songsters and the favourite bird of poets. The curious thing is that they differ so greatly as to whether the Nightingale is a melancholy or a joyous bird. Wordsworth had no doubts at all:

O Nightingale! thou surely art
A creature of a 'fiery heart'...

and lovable seventeenth-century angler and philosopher Izaak Walton thought it miraculous music: 'Lord, what Musick hast thou provided for the Saints in Heaven, when thou affordest bad men such musick on Earth!'

Although most people feel they know the Nightingale, surprisingly few have heard the song in the wild and even fewer manage to see one. It is the cock birds who are the singers and they arrive from Africa in mid-April, the females following ten days later.

Happily for me there is a favoured copse near my Suffolk cottage to which they return each year from Africa. One May morning I waited for hours before I found a Nightingale singing his heart out in a tangle of greenery. Unfortunately, he soon sensed I was there, dropped down into deep undergrowth and was away just like the Nightingale of Keats's poem:

Was it a vision, or a waking dream?
Fled is that music: – Do I wake or sleep?

Great Spotted Woodpecker

Rap, rap, rap, rap, I hear thy knocking bill...

Montgomery

The three members of the woodpecker family that we have in Britain lend a touch of the exotic to our birdlife. Here we see the most common, the Great Spotteds, at their circular nest-hole in an ancient willow. The male is the lower of the two and has a crimson nape; both sexes have red under-tail feathers.

Basil Ede's bird is resting although ever-alert, and supported by its stiff, pointed tail-feathers. The legs are short and the two forward-pointing toes are strong with long curved nails for clinging to the bark. The chisel-like bill is backed by a large strong-skulled head to absorb the shock of continual pounding and probing.

The Great Spotted selects several favourite drumming trees in the nesting area, visiting them in turn. Dry branches are preferred for resonance and they are usually near the tops of trees. Drumming fulfils a similar purpose to bird song: it attracts a female and warns off other males.

When drumming the bird keeps his bill close to the tree and strikes with great rapidity – as many as eight or ten times per second – and the sound can carry about a quarter of a mile. Tunnelling demands a different approach and he has to use the full swing of neck and head. Some American woodpeckers – the flickers – raise the roof by choosing to drum on corrugated iron.

The male Great Spotted does most of the heavy carpentry; digging out the nest-hole may take up to three weeks of hard work. He is extremely wary and at the slightest disturbance will disappear behind the trunk: woodpecker-watching requires great patience and total stillness.

I am lucky because, in winter, Great Spotteds visit the Dougall snack-bar in our tiny Hampstead garden. They cannot resist fat or nuts and a male makes a regular call about lunch time. He swoops down dramatically from a high beech like a demon king, scattering the small birds in all directions. I once invited BBC television cameras into my garden and so the bird became quite a celebrity.

Avocet

How good it is to know that in Britain we are now, at last, making amends to the graceful Avocet, after harrying and persecuting it for so long, especially during the nineteenth century. The bird, which has now become the proud emblem of the Royal Society for the Protection of Birds was forced to give up the struggle in 1842, when it bred on Romney Marsh. A hundred years were to pass before it nested again in this country. The main reason for its failure to breed here was the merciless greed of collectors. Unfortunately the Avocet was greatly prized for its beauty; in addition, the feathers were much in demand by anglers for the making of artificial flies, and the eggs were considered a delicacy. The breech-loaders and punt-gunners were the last straw.

Writing in 1903, the naturalist E. Kay Robinson imagined how the leader of a party of migrant birds, approaching the English coast, might have forewarned his flock: ' "There is England – that ragged, dark line on the horizon: now shut your eyes, and fly for all you are worth!" Bang! Bang! Bang! "Well, how many of us got through this time?" '

The return of the Avocet to this country as a breeding bird has been a triumph for bird protection and, oddly enough, it would probably not have come about at all, had it not been for the war. As a defence measure, large stretches of the Suffolk coastline were flooded and, when the sluices began operating again, it was found that perfect conditions had been created for wading birds. The Avocets were quick to investigate, because they had lost many of their breeding grounds in the Netherlands through wartime action.

It was in the spring of 1947 that four pairs were found to be nesting at Minsmere. The RSPB, which then had only a few thousand members, managed to lease the area as a reserve and also bought Havergate Island in the estuary of the River Alde, where another four pairs were found. Since then, thanks to expert conservation and management, the Avocet colonies in Suffolk have become well established, although, with such demanding birds, constant care and vigilance is required.

Mistle Thrush

The gray thrush heard the thunder's roll,
And sang and heard not what he sang.

Ralph Hodgson

The Missel Thrush, to give the name the old spelling, is larger and grayer than his close relation the Song Thrush and delights in singing during a storm. Of his many names, Stormcock perhaps suits him best of all. Edward Thomas describes how he watched one hurling defiance from the top of a larch: 'There and in oak and ash the Missel Thrush is an embodiment of the north wind, summing it up in the boldness of his form and singing, as a coat of arms sums up a history. Mounted on the plume of the tops of the tall fir, and waving with it, he sings of adventure, and puts a spirit into those who pass under and adds a mile to their pace.'

Not a bird of suburban gardens, he is most easily seen at the end of the breeding season, when quite large family flocks move about over open country, including town parks and playing fields. The young have yellow on the upper parts and are flecked with white.

The Welsh call him grandly 'pen y Llwyn', 'the master of the coppice', and he will readily attack even a Kestrel, when the young or eggs are threatened. In 330 BC Aristotle noticed the bird's liking for mistletoe and other berries and the name has stuck ever since.

Song Thrush

Summer is coming, Summer is coming.
I know it, I know it, I know it.

Alfred Lord Tennyson

May is perhaps the best month of all to hear the Song Thrush, Throstle or Mavis, and Tennyson well knew that he sings his short, clean-cut phrases two, three or even four times over. In contrast with his gentle, retiring nature, the song is a vigorous shout of joy. The bird is also a talented mimic and not only of other birds. In our tiny Hampstead garden my wife and I have been caught out infuriatingly by our local thrush, who has, more than once, had us rushing indoors with his ringing imitation of the telephone.

We find that early morning or evening is the best time to watch him quartering our patch of grass for worms. A hop, a quick rush, then a pause, as he cocks his head for intensive listening – or is it looking – or both? The slightest trace of movement and down shoots his bill for a long, hard pull.

But snails, of course, are his speciality. He finds them under the ivy where they hide from the heat of the day. Sometimes, when he's had all the trouble of smashing them open, I have seen a Blackbird, with devilish timing, dart in and deprive him of his meal. The Song Thrush will have to watch out – it's a hard world!

Red Grouse

The Red Grouse or Moor-cock, (Gor-cock to give it one of its Scottish country names), has the distinction of being unique to the British Isles although from time to time, it has been introduced into western Europe. This true Ancient Briton is essentially a bird of the heather – he feeds on it and nests among it – and seems to embody the very spirit of the high, open moors. For his hardiness, wildness, strength of flight and brilliant aerobatics, sportsmen rate him the king of gamebirds. Lord Home, in his book *Border Reflections*, pays him this tribute: 'The cock bird is proud and possessive, and in his full plumage with the blood-red streak above the eye he looks it, and acts accordingly. Having chosen his nesting site he will defend it against all comers; indeed so tenacious is he that if, by some blight of nature, the food supply around his castle is decimated, he will stay and die rather than move. There are even many authenticated tales of attacks on human beings who have strayed onto his preserve.'

One such experience was reported in *The Field* of 25th November, 1971. A Yorkshire gamekeeper Mr John Coates wrote from Grinton Moor: '...from the heather a cock grouse arose, and set about me. It then clung on to my jacket or stockings, shaking them like a dog with a rat.' Not for nothing has the sound of the cock's call been likened to 'Go-back, go-back'.

For all his gameness, the nesting Red Grouse has a host of natural enemies even on well-keepered moors: crows, stoats, weasels, birds of prey and, worst of all, the fox. An even more deadly threat is a disease caused by the strongyl worm, which multiplies exceedingly when the birds' resistance becomes lowered through a combination of over-population and frosted heather. Unlike Ptarmigan, those birds of the mountain, Red Grouse keep their colourful red-brown plumage all the year round; in heavy snowfalls on the high moors they have to keep treading, so as not to be buried.

At the end of the breeding season, Red Grouse tend to scatter and relate their numbers to the available acreage. This can cause disappointment to shooting-men when 12th August comes round at last and they find the birds' numbers on the moors suddenly much reduced.

Bee-eater

In all the livery decked of summer's pride,
With spots of gold, and purple, azure and green.

John Milton

For the sheer glamour of its multi-coloured plumage and distinguished tail, the Bee-eater has no equal in the British list, but woe-betide any bee or other pollen-seeking insect that flies within reach of that wicked, down-curved bill. Unfortunately we only see Bee-eaters as very rare summer visitors because they are on the extreme edge of their range. These tropical looking birds attract so much attention when they do turn up they get little peace and therefore seldom stay long. In any case they soon reorientate themselves and move off, although in 1955 three pairs stayed to breed in a gravel-pit near Lewes in Sussex. Two of the pairs raised seven young but the third nest was destroyed by curious locals before a guard could be put on it by the RSPB.

In southern Europe and Africa Bee-eaters are common and it is remarkable how, with those long, slender bills and short legs, they are able to dig out nesting-tunnels, up to two-and-a-half metres long. Like Sand Martins, they usually place these in sand-pits or river banks.

My first sight of Bee-eaters was in Majorca. I shall always remember their graceful flight: no bird hunts with a more deadly panache.

Golden Oriole

As with the Bee-eater, the Golden Oriole is on the edge of its range in Britain and this, coupled with the attention its gorgeous livery attracts, results in it rarely staying to breed with us. Here we see a male in all his glory, so we know he must be at least three years old, as the brilliant yellow plumage is not attained until then. The oak is one of his favourite trees, especially if situated at the edge of large woods and fairly close to water.

Owing to past persecution, the bird is so shy in Britain that it seldom stays in view for more than a moment – moving rapidly in undulating flight from cover to cover. The chances are that its melodious, flute-like whistle will be heard before the bird is seen: it seems to know that in the brilliance of its plumage lies peril. In size, it is about the same as a Song Thrush, although wings and tail are longer. The female is duller and greener with the black replaced by dark brown; the young birds are similar.

In southern France, Spain and Italy the Golden Oriole is quite common. I remember my delight at first hearing, then seeing one, in characteristic dipping flight, by the edge of a grove of trees near Roquebrune, the ancient fortified town overlooking the Mediterranean near Menton. There, it did not seem unduly shy: perhaps because it is aware that it does not make good eating.

Swallow

The beautiful swallows, be tender to them.

Richard Jefferies

There is no doubting the Swallow's special relationship with Man: cave-men shared their homes with him, as we do today. I remember the excitement my family felt one May, when the tell-tale small pieces of mud and straw were found sticking to the rafters in a corner of the back porch of our Suffolk cottage.

The trouble was that we were constantly going in and out to the garden. All went well until four young hatched. The parent birds, with much solicitous twittering, carried on bringing food with almost total disregard for us, but reserved their fear and dislike for our amiable, aging Dalmatian, who was relentlessly dive-bombed on all his goings-out and comings-in.

I shall always remember the touching trust the young ones had in us; their enormous gapes and wonderfully expressive eyes, and how they sported chestnut bibs just like their parents.

Sadly, the following summer we were unable to visit the cottage and there must have been interference, perhaps from cats, because the Swallows have not returned.

Another instance of their trust in humans was shown at the nearby RSPB Minsmere reserve. At one time, in each of the eight wooden hides, lodged in the corner as you opened the door, was a Swallow's nest. Hundreds of birdwatchers trudging past daily troubled the occupants not at all.

The Barn Swallow of North America is slightly smaller than our bird, but otherwise almost identical. He spends his winter in South America, while our Swallow migrates over six thousand miles to the Cape.

It is strange to think that in the eighteenth century the great naturalist Gilbert White, musing on where the birds wintered, could have written: 'house-swallows have some strong attachment to water, independent of the matter of food; and though they may not retire into that element, yet they may conceal themselves in the banks of pools and rivers during the uncomfortable months of winter.'

Now, thanks to ringing, we know more about their migration, but still say, as did the ancient Greeks: 'One swallow doesn't make a summer.'

Tufted Duck

And as for the duck, I think God must have smiled
  a bit
Seeing those bright eyes blink on the day He
  fashioned it.
And He's probably laughing still at the sound that
  came out of its bill!

F.W. Harvey

Britain's most common diving duck, the Tufted, is very rare in America and did not breed over here until the nineteenth century. Since then, thanks to our bird protection laws and to the important conservation work of the Wildfowl Trust and the Royal Society for the Protection of Birds, their numbers have greatly increased. The Wildfowlers' Association of Great Britain and Ireland has also helped both in conservation and by ensuring that the number of birds shot is controlled. Even so, the total breeding population in Britain is only four to five thousand pairs and two thousand in Ireland. In winter, our resident birds are joined by big flocks from northern Europe and Iceland; they then spread out to reservoirs, gravel pits and other stretches of inland fresh water.

Once these comic little black-and-white characters have decided to stay and breed, the resultant young show a strong inclination to remain where they are. Lord Grey of Fallodon, who was Foreign Secretary in the twenties, found this inconvenient on the small lake at his home in Northumberland. Day after day, he had to look out at a flock of thirty or so contented Tufties who were crowding out the more exotic varieties. To his embarrassment, they became so tame that they used to waddle up to him and even tug at his shoe-laces, if he disregarded their pleading looks for food.

In the case of Lord Grey, they obviously knew they were on to a good thing: those beady, golden-yellow eyes can size up a situation in a trice. Should they decide you are no use to them, they will take an instantaneous header, slipping out of sight, only to bob up again a few seconds later like a rabbit from a hat.

I'm told that a brood of young Tufties once hatched out in one of the Downing Street gardens and had to be escorted to the lake in St James's Park by one of the Prime Minister's staff – holding up the traffic on the way.

Redstart

The Redstart or Fire Tail is one of Britain's most striking summer visitors but nowadays mainly confines its breeding to the north and west. Here, we see the male in his dramatic finery perched protectively by the chosen nesting site. If you are lucky enough to see it, his courtship is quite spectacular. First, comes a lively chase through the tree-tops; he then flies down to the nest-hole, pops his head in and fans out his flame-like tail in all its glory.

There is one occasion I shall always remember. Having slaved in our Suffolk garden all morning, I was dozing in a deckchair after lunch. The weather had been stormy and I was awakened by heavy rain. Gazing around, still not quite awake, I could hardly believe what I saw: there were Redstarts shivering their tails in the hedgerows all around me, to say nothing of exhausted Wheatears and Wrynecks scattered among the flowerbeds and a host of Pied and Spotted Flycatchers.

The storm had driven them off-course on their journey from Scandinavia and the north to winter quarters in the warm south. Gratefully, they had made a landfall on a twenty-mile stretch of the Suffolk coast. Luckily for me my garden was right in the middle of it, and I was witnessing the biggest fall of migrant birds ever recorded on our coasts. The date was 3rd September 1965.

Spotted Flycatcher

If you scare the Flycatcher away,
No good luck will with you stay.

Anon.

The Spotted Flycatcher is such a modest, mouse-grey, pensive little bird and so wholly beneficial to gardens that I cannot imagine anyone wishing to scare it away.

Here we see him strategically perched on the projecting branch of a buddleia, which is especially favoured by butterflies and the other insects on which he feeds. However thoughtful-looking and dejected he may appear, do not be deceived: this bird is all rapt concentration. The tiniest sound or sight of a winged insect and off he darts in a quick sally of airy action – you may even hear the snap of his long, slender bill – before he returns to the same observation-post to eat his tiny meal, or pause until sure that it is safe to take it to his young.

Often he will operate from the same place for half an hour or so before moving to another convenient branch or post nearby. His voice, little more than a soft, meandering twitter, is as unassuming as his appearance, as if he did not intend it to be heard.

The Spotted Flycatcher is the latest of all our migrants, not arriving from tropical Africa until mid-May, but that of course, is the time when insects are most plentiful. I never tire of watching his intricate evolutions in the air: no summer is complete without him.

Osprey

This is a dramatic study by Basil Ede of the chocolate-and-white fish hawk in perfect harmony with its surroundings. Note the powerful wings with a spread of nearly two metres, the steel-blue toes tipped with long, sharp, down-curved claws, and the pads on the feet with their short spines to grip and hold the slippery prey.

It must seem very strange to Americans that a bird of prey so widespread with them should, in Britain, be national news each spring, when a pair returns from Africa to the tall pine at the RSPB Loch Garten reserve on Speyside.

For us, it is equally strange to hear that in some coastal regions of America where trees are scarce, Ospreys are often reduced to using telegraph poles as sites for their huge nests of sticks, with unhappy results in the disruption of telephone services.

In Britain, on the other hand, there was tremendous interest when in the mid–1950s a pair returned to nest in Scotland after an absence of fifty years. The public's imagination was fired and, thanks to a round-the-clock organisation of watchers set up by the Royal Society for the Protection of Birds, the Ospreys have now succeeded in spreading out to about twenty other secret nesting sites throughout Scotland. This has certainly proved a major conservation victory, although constant vigilance is still necessary during the nesting season. In May 1977 an egg-collector managed to steal the only clutch of four Osprey eggs to be laid in Britain this century; the usual number laid is two or three. So with us the battle is still on.

This tribute to the Osprey comes from an American, Jessie B. Rittenhouse:

> On a gaunt and shattered tree
> By the black cliffs of obsidian
> I saw the nest of the osprey.
> Nothing remained of the tree
> For this lonely eyrie
> Save the undaunted bole
> That cycles of wind had assaulted
> And, clinging still to the bole,
> Tenacious the topmost branches.
> Here to scan all the heavens,
> Nested the osprey.

Osprey—Pandion haliaetus

Green Woodpecker

Laugh, woodpecker, down in the wood...

A.C. Benson

For me, the loud ringing laugh of the Green Woodpecker is one of the most marvellous sounds of summer and I am lucky enough to hear it often, both in Hampstead and Suffolk. The Green is the largest and most brilliantly coloured of the three British woodpeckers. The one in the painting is a male as can be seen by the vivid red stripe under the eye – in the female it is black. The bird has no offering in its bill for the clamouring youngster, as the labour-saving method of feeding is to regurgitate straight into the gullet. In this way a larger amount of food can be conveyed and fewer journeys are required.

The Green also differs from the other woodpeckers in that during the summer he frequently feeds on the ground. A specialist at raiding the nests of ants, he makes good use of his deadly, prehensile tongue, which darts out ten centimetres and is coated with a sticky slime. Once, with binoculars, I saw a bird's tongue being withdrawn from a loose heap in a fir wood. It was black with ants. Thomas Bewick, one of England's most famous natural history artists, wrote of the woodpecker's tongue: 'It has the appearance of a silver ribbon, or rather, from its transparency, a stream of molten glass, and the rapidity with which it is protruded and withdrawn is so great that the eye is dazzled in following the motions: it is flexible in the highest degree.'

Probably because the Green Woodpecker is the most eye-catching of the three and was once the most common, there is a wealth of folklore and legend surrounding him. Apart from the apt name of Yaffle referring to his laugh, he has several others. In the early part of the eighteenth century John Aubrey wrote: 'To this day the country people doe divine of rain by their cry.' Certainly, Rainbird is another of his names, possibly because his laugh is often heard in April, that most showery of months.

Skylark

To me the lark's clear carolling on high
Reveals the whole wide blue, bright sky.

Anon.

The Skylark has always been one of the birds best loved by poets. It does indeed seem to belong more to heaven than to earth being the only British bird which sings while ascending, keeps singing while hovering and continues to sing while descending. He is a firework in the sky but, on the ground, just a nondescript grey-brown bird that runs among the stubble.

Happily, the Skylark is found wherever there is open country, whether it be cultivated land, saltings by the sea, heathery moors or the rolling downs. The bird is now the most widely distributed of any British species, since changes in farming methods with bigger fields and fewer hedgerows have suited it. The Skylark has no need of a songpost to establish his territory – all he needs is air.

How strange that the sun's darling, so celebrated in English poetry, is seen across the Channel as an important sporting bird. In EEC laws on bird protection the French have insisted on the right to continue shooting Skylarks. What makes it worse is that thousands of the larks are British-bred birds moving south in the autumn. It is a blessing that the Skylark is a highly adaptable species well able to hold its own.

Yellowhammer

In early spring, when winds blow chilly cold,
The yellowhammer, trailing grass, will come
To fix a place, and choose an early home,
With yellow breast and head of solid gold.

John Clare

The Yellowhammer, like the Skylark is a bird of open country; in cultivated areas, however, modern farming methods do not suit it so well. It is essentially a bird of the hedgerows, where in early summer the males are a joyous glimpse of bright, clear yellow and reddish-brown.

Unfortunately, around my Suffolk cottage the hedges are fast disappearing to make prairie-like fields for huge machines, and many nesting sites have now been lost. Some areas of heath and common happily still remain, where the Yellowhammer can be seen perched, sentry-like on a topmost spray or sailing with quick undulating flight along the country lanes. Sometimes, when among golden gorse, it is only the distinctive song that helps to spot him: those few bars repeated over and over again bring childhood memories of 'little-bit-of-bread-and-no-chee-eese'.

Unaccountably, in Scotland, the Yellowhammer was at one time held to be the Devil's bird and to be singing 'de'il, de'il, de'il tak ye!' It was even claimed that he drank a drop of Devil's blood every May morning. I prefer the English version and his song for me holds all the drowsy warmth of high summer.

Lesser Spotted Woodpecker

The busy woodpecker
Made stiller by her sound
The inviolable quietness.

Percy Bysshe Shelley

Autumn – and as the leaves turn colour and gradually thin, there is a better chance of catching a glimpse of the smallest and rarest of the three British woodpeckers. He is not much bigger than a sparrow, though infinitely more decorative, and seldom seen away from the southern parts of England and Wales. The bird may not be quite as rare as is sometimes thought, owing to its knack of keeping out of sight in the topmost branches.

Basil Ede's painting shows a male with its distinctive red crown feeding near the top of an old field maple. Any lurking insect will need good fortune to escape that searching, sticky tongue, which has a tip barbed with small filaments like the teeth of a rake. All the woodpecker features can be clearly seen; the long toes, the short legs, the stiff pointed tail-feathers, and the shallow keel of the breast-bone that allows the body to be held close to the trunk.

The white wing-stripes earned it the name of Barred Woodpecker, which in my opinion, is a simpler and more apt designation than Lesser Spotted. Like the Great Spotted it also drums, but the sound is softer and less resonant although the bursts are longer.

I glimpse him occasionally in the old Middlesex Forest area of Hampstead Heath, but he is extremely wary and on the slightest suspicion of being watched, makes off with the characteristic deep, dipping flight of the woodpeckers.

On other occasions he nips round to the far side of the tree and, after hiding there a few moments, his scarlet head will peep cautiously round the trunk again to see if the coast is clear.

I was granted an unusually long look at him recently in my Suffolk garden. It was an early September evening when from the house I spotted a lone bird splashing about like mad in our small stone bird-bath. To my amazement it was a Lesser Spotted male: I had never seen one in the garden before – and I have never seen one since.

Woodcock

A bird of passage gone as soon as found
Now in the moon perhaps, now underground.

Alexander Pope

The Woodcock is the most mysterious of all our game-birds. Charles Morton, the English seventeenth-century ornithologist, was of the opinion that the bird migrated to the moon. He had observed that Woodcocks sometimes landed on to the masts of ships, and this led him to deduce that as they arrived 'right down from above' then they must have come from lunar regions.

The full moon in November is still known as the Woodcock moon and indeed, it often brings falls of migrant birds from Scandinavia and the Baltic region. At about the same time many of our British birds fly to Ireland and some to France and Spain.

Apart from the mysteries of their migration, they are also among the most difficult of all birds to see. In fact the Woodcock is much more likely to see you, thanks to his large eyes near the top of his head with a range of three hundred and sixty degrees. When probing for earthworms with his sensitive bill, he is able to keep a look-out for anything creeping up on him from the rear.

Originally a wader, he has now adapted to life in forests and woodland with plenty of clearings. He often lies up during the day, flighting out to feed in open boggy country at dusk. The best chance of seeing one is during the breeding season when the male makes 'roding' flights at twilight with down-pointing bill. The beats of his rounded, owl-like wings are slow and he repeatedly makes two distinct calls – a frog-like croak and a short shrill chirp – before descending in darkness to join his mate on the nest.

If the young should be threatened before they are able to fend for themselves, the female will sometimes lift them to safety by flying off with them clutched between her thighs.

Even gun dogs find them strange: I heard of one young retriever, who cautiously approached his first Woodcock, dead in the heather, and promptly cocked his leg on it.

Coal Tit

This tiny son of life; this spright...

Walter de la Mare

The smallest of the British titmice is also the perkiest and, although by reputation shy, I would never have guessed it from watching the members of his tribe which come to my Hampstead garden in late autumn and winter. They are well able to take care of themselves among the host of other small birds attracted to the hanging food-containers. My Coal Tits also show considerable prudence by storing food – flitting tirelessly between the containers and their hoarding places – in crevices located in the support-posts of a nearby pergola. So engrossed do they become in this activity that I can approach to within a metre or so of them and they show no fear.

As we can see in the painting, there is no difficulty in identifying these little birds as the broad white patch on the nape is clearly seen. At this time of year when the breeding is over, Coal Tit families stay together and often form foraging parties in mixed woods with other tits, Goldcrests and Treecreepers, moving from tree to tree in ceaseless activity.

Sometimes, in rough winter weather among the birch woods, I have seen what at first glance appear to be tiny black fruits dangling from the ends of spindly boughs, tossing in the wind. But in no time, the fruits multiply as other tits join the merry company; then, chuckling and calling to each other, they scatter far and wide.

Long-tailed Tit

A troop of birds on laughing wings
Came tumbling by in loops and strings:
'See! See! See!' their leader cried...

Anthony Rye

At about this time of year family parties of Longtails roam happily through the woods. For me, Anthony Rye's words conjure up perfectly the sight and sound of their high, thin calls, like tiny electric bells 'see-see-see'. F.O. Morris, the nineteenth-century naturalist, described the sound as the very embodiment of gentleness, weakness and tenderness.

Were it not for the long tail and the puffed-out plumage, this diminutive creature could claim from the Goldcrest the distinction of being Britain's smallest bird.

A distinction it can claim is that of constructing the most marvellous nest, surely the most perfect of all nests built by British birds? In shape it is oval with a small entrance-hole near the top. The principal materials used are lichens, moss, wool, spiders' webs and, above all, feathers. The interior is cushioned right up to the entrance with softness, nearly three centimetres deep, and as many as three thousand individual feathers have sometimes been used. No wonder one of the country names for the Long-tail is Feather-poke.

The tail of the sitting bird, mainly the hen, is held above her back; her head and the end of her tail then serve to cork up the entrance. That would account for two more names: Bottle Tit and, rather unkindly, Bum Barrel.

House Martin

The House Martin is a close relation of the Swallow and sometimes confused with it although slightly smaller in size. Basil Ede's pair clearly shows the other distinguishing features: the white underparts, feathered right down to the toes, the distinctive white rump, and the lack of long, outer tail-feathers.

House Martins arrive in this country a little later than Swallows. In the air they appear black-and-white and tend to fly higher and they are even more devoted to Man's dwellings, sometimes following him into large towns. As insect-eating birds, they have gained immensely from the Clean Air Acts and have gladly returned to inner-city life.

They also differ from Swallows in their custom of nesting in colonies, especially in the country and near water. On June mornings what a delight it is to watch the House Martins gathering at pond edges to collect the mud for their building work; and what fastidious care they take to keep their white underneaths clean, wings raised high, as a girl might pull up her skirt when paddling.

The nests are placed just under the eaves, or sometimes in window-corners for extra protection, and they are marvels of architectural and plastering skill. Unfortunately, sparrows very often take over these fine structures, placing in them their straggling nests of straw. One of the especially pleasing things about being favoured with Martins under the eaves is the companionable, low, bubbling sound of their warbling.

Each autumn around the coast there is the moving spectacle of House Martins gathering for migration. Telephone wires, church towers and barn roofs are thronged with tiny figures, twittering excitedly as they prepare for their mysterious journey across thousands of miles of land and ocean. Walter de la Mare found marvellous words to describe this as he once stood gazing up at the 'ardent legion wild for flight':

> Each preened and sleeked an arrowlike wing;
> Their eager throats with lapsing cries
> Praising whatever fate might bring –
> Cold wave, or Africa's paradise.
> Unventured, trackless leagues of air;
> England's sweet summer narrowing on;
> Her lovely pastures: nought they care –
> Only this ardour to be gone.

Blackbird

The ousel-cock so black of hue
With orange-tawny bill...

William Shakespeare

Here we see the mellowest singer of them all bathed in golden, autumn sunshine, and who would begrudge him a peck or two of those succulent looking apples? He has a wary look in that yellow-rimmed eye and yet seems perfectly relaxed, as if he knows he will come to no harm. This represents a complete change in the Blackbird's behaviour, because until about the middle of the nineteenth century he was a shy bird of the woods. Fortunately, he has been able to adapt to changed conditions, acquire confidence in Man, and today in Britain has become one of our most numerous and widespread songbirds.

Luckily for him, he is able to eat a very varied diet and there is no doubt that food provided at bird-tables in hard weather has helped immensely in his successful adaptation to city life. Fruit of all kinds he finds irresistible, but some gardeners may feel he pays for all he takes with the wondrous beauty of his song. Another of the most important items on the Blackbird's diet sheet is the earthworm and he sets about the job of finding it with the utmost application. He will also take insects, spiders, slugs and even wasps. I once caught one in the act of trying to catch a minnow in my tiny garden pond.

At this time of year, Blackie is often heard rustling about under bushes in his search for insects among the piles of dead leaves. It is this ability to eat almost anything that helps him survive severe winter weather more successfully than his cousin the Song Thrush. A hen Blackbird is sometimes mistaken for a Song Thrush, as her colour is brownish with a pale throat and her underparts are mottled and spotted. There could never be any mistaking the cock, however, or his joyous song as T.E. Brown noted:

Just listen to the blackbird – what a note
The creature has! God bless his happy throat!
He is so absolutely glad
I fear he will go mad.

Kestrel

My feet are locked upon the rough bark.
It took the whole of Creation
To produce my foot, my each feather:
Now I hold creation in my foot.

Or fly up, and revolve it all slowly –
I kill where I please because it is all mine.
There is no sophistry in my body:
My manners are tearing off heads –

Ted Hughes

The Kestrel is the one bird of prey still widely seen in Britain and its numbers have been estimated at one hundred thousand. The bird has become a great attraction for drivers on the motorways and for many youngsters Kestrel counting has become a popular pastime. It is no wonder that in 1965 the Windhover, as the Kestrel is sometimes known, was chosen as the badge of the junior branch of the Royal Society for the Protection of Birds and the little falcon must have won the club a great number of new members. It is a measure of the bird's adaptability that he was so quick to realise the advantages of exploring and hunting over the grass verges created by the motorways. With no agricultural sprays and little disturbance these form important reserves where small mammals and birds can flourish.

His move into the cities did not begin until the 1930s and was probably due to overcrowding in the country, where gamekeepers were at last realising that the Kestrel, as a destroyer of rodents, was an ally rather than a foe. In no time the bird was colonising central London and other big cities, where high buildings offered plentiful nesting sites and sparrows provided abundant prey.

Two years ago, Kestrels nested in a tall lime close to my Hampstead cottage. It was fascinating to watch the handsomely marked male gliding back to his tree-top family, to the accompaniment of excited squeals from the impatient chicks.

BASIL EDE © 1976

Treecreeper

Mouse up a tree,
White belly
Close to bark,
Curved bill
And tail support.

Robin Ivy

A young birdwatcher-poet neatly complements Basil Ede's delicate painting. The Treecreeper is one of our smallest birds and one of the most difficult to see, although it stays with us all the year.

The best chance of seeing one is in late autumn when the leaves are down; you might then catch a fleeting glimpse of his silvery underneath. Watch closely and you will almost certainly see him reappear, as he corkscrews up the trunk in a series of short jerks or shuffles. He is forever probing with that slender, scimitar-shaped bill, meticulously searching for tiny spiders, woodlice and other minute insects. On nearing the higher branches, he drops away slantwise to the base of another bole and starts his spiralling movement all over again.

Anthony Rye's description of a Treecreeper working a tree in winter is marvellously evocative:

As if brown, decaying leaf
Back to the branch were blowing.

Bluethroat

In Britain the Bluethroat is a rarity and the sight of one would make a red-letter day for any birdwatcher, especially when set as here, against the golden autumn foliage of the sweet chestnut.

This is the time when numbers of them pass down the east coast from Scandinavia on their way to winter quarters in north-east Africa, but as yet, I have not had the good fortune to see one. The Bluethroat belongs to the family of small chat-like birds and is related to the Robin, Whinchat, Stonechat, Redstart and Nightingale.

Like our Redbreast he has a habit of spreading out his tail and flirting it up and down and, as befits a relation of the Nightingale, he is also said to have a fine song. Miss Haviland, a nineteenth-century naturalist, described it lyrically, likening the sound to 'a golden pea', which 'leaped and vibrated in the pipe behind his gaudy bib.'

The female, in comparison, is fairly drab. She has a whitish bib, bordered by streaks of dark brown and her underparts are a pale buff colour. It is known that a pair nested in Scotland in 1968.

My greatest opportunity of seeing a Bluethroat came on 3rd September 1965, when easterly winds blew a number of them on to the Suffolk coastline in the region of my cottage; but somehow I missed them.

Pheasant

See! from the brake the whirring Pheasant springs,
To mount exulting on triumphant wings;
Ah! what avail his glossy varying dyes,
His purple crest and scarlet-circled eyes,
The vivid green his shining plumes unfold,
His painted wings, and breast that flames with gold.

Alexander Pope

A wild cock Pheasant, glinting like a sultan in pale autumn sunshine is one of the most splendid sights in the British bird world. However, in Suffolk at this time of year the birds are so numerous that they seldom get a second glance as thousands of reared Pheasants are released for the shooting season which starts on 1st October.

I do not shoot, but am nevertheless able to recognise that many of our songbirds can find food and nesting sites thanks to the rough areas left as coverts for Pheasants. In hard winters many small birds are also undoubtedly saved by eating the seed scattered for Pheasants at regular sites by gamekeepers.

On the negative side, birds of prey are often needlessly and illegally persecuted by keepers. As a result, sadly, in those parts of the country where the shooting of game is big business, fewer hook-beaked birds survive. There are, of course, many arguments for and against shooting, but on balance, it would seem to me that sensibly managed game preserves can be an aid to conservation.

It is probable that the Romans introduced the Pheasant to Britain, although the first documented evidence is a bill of fare belonging to Waltham Abbey in Essex, which first had the bird on the menu for the monks there in 1059.

The name itself derives from the river Phasus in the ancient province of Colchis to the east of the Black Sea. Another form of Pheasant with a white neck-ring was introduced in the eighteenth century from eastern Asia and there are now many hybrids, some with and some without the neck-ring.

In spring the males become most pugnacious, hacking away viciously with their spurs like game-cocks; and the sudden, explosive sound of their crowing is then one of the most exciting sounds of the woodland or reedbed.

Tawny Owl

What music this – high and remote,
Fleeting above the frozen earth?
Vision-translated wreaths that float
Of wood-smoke undulant blown forth!
What lovely sound! What subtle birth!
Music attenuate, tremulous, clear...

Anthony Rye

The Tawny or Wood Owl is a bird we are far more likely to hear than see as it is essentially a creature of the night. From late August onwards, as soon as the parents have chased off their young to another stretch of the wood, they are at their noisiest. That is when my wife and I are often wakened in our Hampstead cottage by the shrill 'kee-wicks' and the wavering 'hoo-oo-oo', as the Tawnies resume their courtship and territorial squabbles. In the middle of the night it is an eerie sound, as though from another world, and yet welcome.

Like most birds, owls cast pellets of undigested food thus disposing of the bones, fur and feathers of the small rodents, beetles or occasional birds on which they prey. In the garden I frequently find these five centimetre-long, greyish-green, cylindrical objects round the base of a tall beech tree used as a daytime roost.

Basil Ede's owl is evidently just meditating and waiting for darkness to fall, before he glides away on soft, silent wings to locate and drop on an unsuspecting vole in the tussocky grass. Although well-nigh invisible from the ground, as he sits bolt upright, close to the trunk of the tree, he is by no means as sleepy as he looks. The huge eyes, which are bigger than a human's, are fixed in their sockets, but the owl is able to turn his head full circle – and so misses nothing. His ears too are exceptionally large, though concealed by a skin-flap and strangely the right one is slightly larger than the left. This may well help him to pinpoint the tiniest rustling sound made by a small mammal in the dark.

The owl is wise to stay out of sight: once small birds spot him they gang up and give him hell. So, when in a wood you hear a tremendous hullabaloo – watch out! It may be your best chance of seeing a Tawny.

Pintail

Basil Ede has given us a superb study of that aristocrat among ducks – the elegant Pintail with his tail-streamers fully twenty centimetres long. In Britain we know him principally as a winter visitor, especially to estuaries in the north-west of England and to that wildfowl paradise, straddling the borders of Cambridgeshire and Norfolk – the Ouse Washes. Here, in the heart of the fen country, we have one of the most important wintering places for wild swans and ducks in the whole of Europe.

The Pintails' main breeding grounds are in the far north in Iceland and also in northern and central Russia and parts of Siberia. Unfortunately, over the whole of the British Isles, in most years there are no more than about twenty breeding pairs. How different from North America where they are among the most abundant of ducks. In Europe too they are quite common. Fortunately they are strong, fast flyers, which is just as well for their flesh is said to be tasty. The country name Cracker may stem from the male's croak in courtship.

The Pintail or Sea Pheasant is one of the dabblers and like the Mallard, gets much of his food by up-ending, although his longer neck gives him the advantage of being able to feed in deeper water. The female is dowdy in comparison – a uniform dullish-brown – and having a slightly shorter neck she is restricted to shallower water when feeding. Her drabness can, of course, be very much to her advantage when nesting, as it enables her to merge imperceptibly into the background.

One of the strange things about Pintails is their extraordinary knack of divining the exact moment when the low lands are flooding: within forty-eight hours of floods developing on the washes of East Anglia, I have known flocks take off from Kent and more distant places to converge on the new feeding grounds. This opportunism is of great advantage to the Pintail, as these southerly regions to which he moves in winter are liable to dry up; and such mobility can be his salvation. It is always a great thrill to watch the first of the Pintails dropping in on the washes – sporting and playing, as though celebrating their arrival – streamlined and white-breasted among the dense packs of Wigeon.

Nuthatch

How strange it is that this cheery, pugnacious little bird – compact and all-of-a-piece as though sculptured – should be so little celebrated in literature.

London's Hampstead Heath is still blessed with much old timber, so the Nuthatch, like the Great Spotted Woodpecker, is frequently seen at the Dougall bird-table each winter. He comes to the hanging containers of fat and nuts in direct, horizontal flight like a tiny, aerial torpedo and might land any way up.

In the woods he is difficult to spot, unless you home-in on the sound of his irrepressible, bubbling song. He is no great musician, but has a range of liquid-sounding whistles, which ring out so cheekily that many a girl has suspected an admiring wolf-whistle. The sound has also been likened to the noise made by a pebble, thrown bounding over the ice.

In the painting, we see him, a tiny, statuesque figure, in ideal conditions of luminous, autumn sunshine. But not for long. He has urgent business afoot seeking out minute insects and spiders in crevices of bark. I always find it endlessly fascinating to watch him in erratic, haphazard, gravity-defying movement. There is no limit to what he can do; he is just as happy scuttling mouse-like down the trunk of a tree as up it, and all this is done with the normal perching-foot of three strong, clasping claws pointing forward and one behind.

The Nuthatch also shows considerable ingenuity when nesting. A favoured site is the old hole of a woodpecker. To prevent a take-over by other birds, especially Starlings who are merciless rivals, he meticulously plasters round the entrance with mud, until it is too small for them to penetrate. Heinz Sielmann, an authority on the climbing birds, once watched a running battle in which, as soon as the Nuthatch had finished his plastering, a Starling would try to break in by removing pieces of wet mud in in his bill. Day after day the Nuthatch carried on building up his defences, until enough of the mud had dried into an impregnable hardness, and his little family was finally completely secure.

Wren

Come and make your offering
To the smallest, yet the King.

Anon.

The mouse-like Wren is known surprisingly in legend as the king of the birds. He won the title by flying even higher than the eagle. All he did was hide under the eagle's feathers, and when it had finally soared above all the other birds, out popped the little Wren, flew just above him, and won the crown.

Although traditionally held in great affection, Jenny Wren, Stumpy or Our Lady's Hen, to give just a few of its names, was until recently the victim each year of a barbarous ritual on Boxing Day. According to folk legend, a Wren had betrayed St Stephen, the English martyr, by alerting a guard and preventing the saint's escape from imprisonment. So, on 26th December, St Stephen's Day, bands of youths used to take part in a ritual Wren hunt, beating the hedgerows and killing any Wren they found.

Not everyone realises that our Wren originally came from the Americas. It is not known exactly when he came to Europe, but it was a long time ago, because remains have been found in Great Ice Age deposits. We do know that he is the only member of a large New World family of fifty-nine species to have done so. There, the bird is known as the Winter Wren and is found from Canada right down to Georgia. With us, he is one of the most widespread of all British birds, thanks to his ability to exist in all manner of conditions, in fact wherever there is a neglected corner with spiders, earwigs, wood-lice and other creepy-crawlies to be found.

This tiny, independent dwarf of the woods has an exceptionally loud trilling song, a burst of sweet music which can be heard even in mid-winter. Nevertheless, in very severe weather, Wrens have little to sing about and suffer great problems in keeping warm. In the coldest days of 1979, I even heard of thirty-seven Wrens being found at Frome in Somerset roosting or bundling in a nesting-box measuring only 7·5 × 7·5 × 15 centimetres. Close quarters, but cosy!

Fieldfare

One hears the whirring flocks of Fieldfares sweep
With harsh exultant clamour as they glean
Their beggarly harvest...

F. Brett Young

In Britain the Fieldfare signifies frost in the air and snow on the
ground; in fact the Shropshire name for him is Snowbird and that is
how we see him here in Basil Ede's striking painting. The blue-grey
head and rump distinguish him from his close relation the Mistle
Thrush, and this dandy of all the thrushes cuts a dashing figure when
all around is cold and grey.

In October and early November, with the wind from the east, small
flocks come in over the east coast from their breeding grounds in
northern Europe and range over rough, open country in search of the
winter berries they seem to prefer to their more usual worms and
insect food. It is then that we hear their harsh chacking note. On the
ground the Fieldfare flocks are shy and wary. They stand erect with
heads upraised and, when feeding, all move in the same direction. If
alarmed, they fly high up into trees, perching with their heads into the
wind.

The birds were well known to our nineteenth-century countryman
poet John Clare:

...flocking Fieldfares, speckled like the thrush,
Picking the red haw from the sweeing bush
That come and go on winter's chilling wing
And seem to share no sympathy with Spring.

Redwing

If the Fieldfare is the northern European equivalent of our Mistle Thrush, so the Redwing is of the Song Thrush. The Redwing flocks also come in over the east coast on October nights and their soft flight-calls, 'see-ip' can often be heard high up in the night sky. These shy, gentle birds are even more vulnerable in hard weather than Fieldfares, as worms and insects are an essential part of their diet. In the prolonged freeze-up of 1963 they suffered greatly and were driven by desperation into suburban gardens, as were Fieldfares and Bramblings. The Redwings in particular looked sad, as they hung around disconsolately, as if resigned and waiting to die. Certainly thousands perished that terrible winter.

Although there have only been isolated cases of Fieldfares breeding in Britain, Redwings are increasingly reported to be nesting in Scotland and perhaps, one day, the tender Norwegian Nightingale will finally decide to try breeding regularly in the south. In his *Wild Life in a Southern County* Richard Jefferies describes how one spring he heard a bird in full song in a Surrey oakwood:' – this Redwing was singing – sweet and very loud, far louder than the old, familiar notes of the thrush. The note rang out clear and high, and somehow sounded strangely unfamiliar among English meadows and English oaks.'

White-fronted Goose

The White-front is the most handsome and most easily identifiable of the five grey geese we see in Britain thanks to the distinguishing white forehead and the black bars on his belly. All of them are winter visitors, the Greylag being the only one to breed here.

Perhaps it takes the best kind of wildfowler-conservationist to appreciate these birds to the full; this tribute was paid under the pseudonym of 'BB': 'Wild geese are such noble creatures. Think of their hardihood. On the wild winter nights, when the great winds are up and out, and the rain drives like steel knives horizontally over the flats, the geese do not shelter like other animals. The ducks seek out some sheltered bank or pool set in the hills, every living creature seems to creep into some hole or cranny. But these splendid birds face out the stormy nights, head into wind, unafraid, hardy as old thorn trees.'

White-fronts are certainly the finest flyers of them all and masters of manoeuvre. The best place to see them in Britain is on the saltmarsh known as the Dumbles of the Slimbridge headquarters of the Wildfowl Trust on the Severn Estuary. It is a breath-taking sight to see them spiralling down in a 'whiffling' dive with wings half-folded. Their numbers build up from mid-October to January and February, when there may be as many as five thousand. Most of them come from Arctic Russia and have only short sea crossings over the Baltic and the North Sea.

Another race of White-fronts comes from breeding-grounds in Greenland, navigating over hundreds of miles of ocean to head for a boggy area in south-east Ireland, known as the Wexford Slobs. There was considerable concern among conservationists in the late 1960s when the district was drained for agriculture; but the White-front just gave a hysterical cackle and got on with the job of adapting to the changed conditions. I am glad to say he is thriving; after all, one of his country names is Laughing Goose.

White-fronted Goose
Anser albifrons © BASIL EDE 1977

Siskin

No bird whose feathers gaily flaunt
Delights in cage to bide.

Shackerley Marmion

Until the Bird Protection Acts, the Siskin all too often ended up in a cage. Today, fortunately, Siskins in Britain are thriving as never before. In the painting we see an elegant cock on a sprig of Norway spruce. It is because of the immense growth of Forestry Commission coniferous woodlands in which they breed, that Siskins have now spread to southern districts where they were formerly almost unknown. They have even taken to visiting suburban bird-tables, lured by the hanging plastic bags of peanuts, and it has been noted that they have a strong preference for the bags coloured red.

Close to my Hampstead cottage there are groves of birch trees, on the seeds of which the little birds like to feast, and it is a special thrill in winter to glimpse a mixed flock of Siskins and Redpolls flitting joyously from tree to tree. As the American nature poet Theodore Roethke put it:

And the small mad Siskins flit by
Flying upward in little skips and erratic leaps...

I have often felt as he did, when he ended his poem:

And I seem to lean forward,
As my eyes follow after
Their sunlit leaping.

Waxwing

Many a suburban birdwatcher, on seeing a small group of exotic characters like this in his garden in mid-winter, has been convinced that they were escapees from some tropical aviary. In fact Waxwings breed in sub-Arctic fir woods from Scandinavia to Siberia. At irregular intervals, they migrate, or rather irrupt to this country in late autumn, sometimes staying until April.

Their erratic movements appear to be mostly governed by the state of the local rowan crop, as they depend largely on the berries for their winter feeding. Given a mild winter and plenty of berries at the breeding grounds, the Waxwings will stay put, but when their numbers have built up and the berry crop fails, then over they come. These 'Waxwing winters' have been recorded in Britain ever since 1679.

I remember visiting Newcastle-upon-Tyne in November 1965 and finding the city parks and gardens full of these sleek strangers gorging themselves on the ornamental trees and cotoneaster shrubs. Their tameness was remarkable, perhaps because there are few domestic cats in the northern pine woods. They were paying no attention to city traffic or passers-by and it was fascinating to watch them.

The name Waxwing comes from the curious scarlet 'wax' tips on some of the secondary feathers. Other names for the bird are Bohemian or Wandering Waxwing and, in America, it is sometimes called the Carolina Chatterer.

Goldeneye

He of white-cheeks, and Golden-eyed,
Of white and black alternate pied.

Bishop Mant

Now you see it – now you don't: the buffle-headed Goldeneye is an expert diver and vanishes from the surface of the water at the slightest alarm, being able to stay submerged for up to thirty seconds. The size of his head is part of the secret of his diving ability. Behind the nostrils there is a large air-space linked to the sinuses. It is this which causes the pronounced hump on the head and which is thought to provide him with a reservoir of air when diving. I have heard it said that, if a Goldeneye should be pursued by a bird of prey when in flight, he can shoot straight down into the water and disappear.

He is a swift flyer and the sound he makes on the wing accounts for two of his country names: Rattle-wings and Whistler. T.A. Coward claimed to have heard the whistle of a drake's wings half-a-mile away and likened the noise to thin ice cracking under the bow of a boat.

At first glance, the drake appears black-and-white and the round white patch behind the bill is most conspicuous, even in flight. If he allows you to get close up to him, the large head is seen to have a dark green metallic sheen which provides a dramatic setting for that jewel-like, golden eye. The female is brownish-grey with a chocolate-brown head and lacks the distinctive white spot.

In Britain, the Goldeneye is a winter visitor from its breeding grounds in the forests of Scandinavia and the Soviet Union, where it nests in holes in trees. These are often in short supply, and so in Scandinavia, nest-boxes are provided and readily accepted. In fact Goldeneye 'farming' has become a cottage industry there: the down and eggs are taken three times before the ducks are allowed to brood.

The Royal Society for the Protection of Birds has also provided nest-boxes on some Scottish reserves and in most years several pairs have bred with varying success.

Goldcrest

And the tiny gold-crest, like a flittermouse
Cheeps in the swarthy cedar's topmost boughs.

F. Brett Young

The Goldcrest is one of Britain's two smallest birds: five of them together would weigh only twenty-eight grams, or one ounce, and without all those feathers they would be little bigger than bumble-bees. Here, the male is the lower of the two; and, as they seldom leave the shelter of hedges and trees, it is often only the topknot, glowing like a tiny orange beacon, that makes it possible to see them at all.

Goldcrests are found wherever there are conifers: a recent estimate put their numbers as high as three million; but in spite of this, they are often overlooked. Incidentally, the bird hasn't really got a crest at all, just a bright strip of feathers extending from near the bill to the back of the head, but this can be raised slightly during the mating display.

I like W.H. Hudson's description of the tiny creature's love-dance: 'It hovers on rapidly-vibrating wings, the body in almost a vertical position, but the head bent sharply down, the eyes being fixed on the bird beneath, while the wide-open crest shines in the sun like a crown or shield of fiery yellow.'

Spare a thought for Goldcrests in hard winters. In some years they have almost been wiped out, but fortunately have always managed to build up their numbers.

Snipe

So lonely his plaint by the motionless reed,
It sounds like an omen or tale of the dead.

James Hogg

Outside the breeding season, the Snipe is a secretive solitary bird. As seen here in the grip of winter, when unable to probe for the creeping things which live in mud, he must move on to a milder district or die. His markings help him to merge almost indistinguishably with the reeds at the edge of the frozen mere. All you may see of him is a momentary glimpse of white chest as he explodes into the wind with a harsh cry and sudden zig-zag flight.

In spring and summer there is a better chance of spotting him. Your attention may first be caught by a sound rather like a kid bleating in the sky – the so-called 'drumming' of the Snipe. Look up and you may see the bird plane down from a considerable height, his two outside tail-feathers at an angle of forty-five degrees, vibrating in the rush of air to produce this unforgettable sound. Not surprisingly, a Scottish country name for him is Heather Bleater and Robert Burns called him the Blitter frae the boggie.

In the baking summer of 1976 a Snipe stayed in my garden in Suffolk for the best part of a day; and in winter I have even seen one zig-zagging away from a reedy pond on London's Hampstead Heath.

Teal

The Green-winged Teal is the smallest of our ducks and, I think, the nattiest. What could be smarter than the drake's glowing chestnut head and distinctive dark green eye-stripe? Yet, in the field, the most conspicuous mark is often the long white line on the wing contrasting with the black line below it.

His brown, speckled mate is even smaller and more susceptible to cold, which accounts for the earlier movements of females to warmer areas in the south. As breeding birds, Teal are most numerous in Scotland, Ireland and the north of England, although they are greatly outnumbered by flocks of winter visitors from Iceland and northern Europe.

The nests may be among heather, tussocky grass or even in a wood; sometimes it is a mystery how the tiny ducklings ever manage to get down to the water but Teal are diligent, resourceful parents, and some naturalists believe that, like Woodcock, they are even capable of carrying their young. Henry Douglas-Home, for one, is almost convinced. In his book *The Birdman* he wrote: 'This particular bird had chosen to nest a few yards inside a young moorland plantation of scotch firs. The trees were surrounded by a new, stoutly stobbed wire fence, and since the wire mesh was far too fine for even a Teal duckling to squeeze through and far too tight at the bottom to be burrowed under, we wondered how on earth the female was going to extricate her brood. Of course she managed, the whole family duly appearing on a nearby loch, but not when we were there to record it, so the mystery remained. I see no way in which she could have got the birds out without holding them between her thighs and flying each one over in turn.'

Outside the breeding season Teal tend to stay together in compact flocks and, if put up when feeding, they take off vertically at great speed, dashing away in turning and twisting flight like waders. With good reason is the collective name for them a 'spring' of Teal.

Firecrest

The beautifully marked Firecrest and its close relative the Goldcrest are often confused: Basil Ede's pair of Fire-crested Kinglets shows clearly the main distinguishing feature – the black-and-white eye-stripe. Here, the male is on the left; his crest has more flame to it than his mate and, in general, his plumage is somewhat brighter.

Up to a few decades ago, Firecrests were thought of simply as occasional winter visitors from Europe: there was no question of their breeding in the British Isles, but since then, they have begun spreading across Europe in a north-westerly direction. So much so, that in 1961 the first pair was reported breeding in the New Forest area and succeeded in bringing off young. A gap of ten years followed until the next recorded case of breeding. Then, after a census carried out by the British Trust for Ornithology in 1972, it became clear that the little birds were beginning to colonise southern England in quite a big way, and there was also a breeding report from Lancashire.

Firecrests are said to be highly sensitive to interference when nest-building; for this reason, they were given special protection under Schedule I of the Protection of Birds Act.

In general, these diminutive birds favour conifer plantations and like the Siskin have doubtless benefited by the spread of commercial planting undertaken by the Forestry Commission. This is one compensation for damage sometimes caused to other forms of birdlife, when large-scale planting is undertaken. The nest itself is said to be like that of the Goldcrest and, similarly, is often suspended like a tiny hammock under the branch of a fir tree. It is a marvellous piece of work fashioned out of mosses, held together with cobwebs, and thickly lined with feathers.

What a bonus for birdlovers that this beautiful little bird should have become England's smallest colonist.

Robin

Art thou the bird whom man loves best,
The proud bird with the scarlet breast,
Our little English robin?

William Wordsworth

The Redbreast has long enjoyed a special relationship with Man: even when it was common practice for songbirds to be trapped for eating or sold as cage-birds, he was victimised very much less than most. As William Blake put it:

A Robin Redbreast in a Cage
Puts all Heaven in a Rage.

Wordsworth, when he called him 'the pious bird', was remembering one of the many folk tales about the origin of the red breast. The Robin was said to have plucked thorns from Christ's crown at the crucifixion and, in doing so, received wounds himself. Legend also maintains that he would charitably cover with moss and leaves any dead bodies he found. This probably arose from his custom, as a woodland bird, of closely following travellers who passed through his domain.

Although perfectly able to survive in the woods on his own, the development over the centuries of suburban parks and gardens came as a heaven-sent new habitat to which he speedily adapted. As a result, the Robin has become Britain's best-loved bird and the friend of every gardener. There is nothing he likes better than supervising the digging, as John Clare once observed:

Up on the ditcher's spade thou'lt hop
  For grubs and wreathing worms to search;
Where woodmen in the forest chop,
  Thou'lt fearless on their faggots perch...

There is no greater enthusiast at countless bird-tables, where he is happy to investigate almost anything put out for him; cheese and currants are great favourites and he'd sell his soul for a mealworm. Do remember, however, that there are cats about and it is not always in Robbie's best interest to make him too trusting and tame.

In North America the early British settlers felt very much more at home when they found almost everywhere a thrush with a red breast. They promptly called it a Robin and, although larger than our bird, Americans still know it by that name today.

Grey Heron

Ye fisher herons watching eels.

Robert Burns

Here we see the Grey Heron with eel-like neck and stabbing bill at the time of his greatest danger in mid-winter. There is good reason for the anxious look in that sharp, round eye; if those shallows ice over, he knows it will mean a starvation diet. Severe weather is the bird's chief enemy and after the winter of 1963 the population was reduced to about two thousand pairs. Recovery was slow but, in general, the outlook for Old Franky, as he is often known in Suffolk, has never been so bright. This Suffolk name suits him well because he has been around for millions of years, and 'Frank' resembles the sound of his harsh cry.

The ancient, ungainly bird is one of the great survivors: fossilised bones, dating back to 10,000 BC have been found in Clevedon Cave, Somerset and also in Derbyshire's Pin Hole Cave. Yet, until comparatively recent times, Herons have been hunted for sport and food. They were much used in falconry in the Middle Ages and are the subject of many a quaint tale.

Writing in 1635, John Swan maintained that if the Heron rose above the hawk 'then with his dung he defiled the hawk, rotting and putrifying his feathers'. The last time a Grey Heron was hawked in England was at Didlington in Norfolk in 1838. A Heron will always soar to safety when attacked, which is one of the reasons it was so prized by falconers.

At one time the nestlings were thought to be excellent eating and as recently as 1812 six birds were among the roasts at a feast in the Hall of the Stationers Company in the City of London. The Grey Heron no longer appears on the menu at royal banquets, however, and so the future for Old Franky seems assured. There is even a small colony of wild Grey Herons on one of the islands in London's Regent's Park. Goldfish soon began vanishing from garden ponds nearby; a man I knew found that a stone heron placed in his pool helped keep the live marauders at bay.

Bearded Tit

Basil Ede's pair of beardies are seen here straddling the reeds – they can even do the splits. From the wicket-gate at the back of my Suffolk cottage, in five minutes I can wander down to the edge of the largest area of reedbed remaining in Britain. In area, together with the expanse on the nearby RSPB Minsmere reserve it amounts to about a thousand acres. This is the principal stronghold of that uniquely charming little bird misnamed the Bearded Tit. The male, on the left, has a mandarin-style moustachial stripe and does not even belong to the titmouse family, his nearest relations being the parrotbills of Burma and China.

The couple in the painting have an apprehensive look. They suffer considerably in prolonged periods of hard weather and after the winter of 1947 only four pairs were recorded at Minsmere. T.A. Coward tells us that to keep warm the male and female roost side by side, snuggling together on the same stem, and that the cock often shelters the hen under his wing. That I would dearly love to see.

Fortunately, these sprites of the reedbeds are prolific breeders and in normal summers they have three broods of four to six young, so their numbers soon recover. Nevertheless, in the nineteenth century, owing to intensive drainage of marshes, it seemed they might well become extinct in Britain. As they became rarer, so the demand for them increased. Collectors sought them for museums, taxidermists put them in glass cases and others were incarcerated in cages. Happily, legislation helped to save them. Now, thanks to protection, conservation and new areas of reedbed in East Anglia, their numbers are higher than at any time since records were kept.

I know of few pleasanter sights than to glimpse one of their family parties moving over and through the reeds in late summer. First, one hears metallic 'ching-ching' sounds as they call to each other; then, with luck – there are the beardies – perching like tiny gymnasts on the slender stems, swaying in the wind.

Bearded Tit ~ Panurus biarmicus   © BASIL EDE 1977

Black Redstart

Like his close relation, the common Redstart, this character also has a fiery tail, which he is constantly flirting. The name goes back to Anglo-Saxon times, 'steort' being the Old English for tail. The birds have little in common when it comes to the choice of a nesting place: instead of the edges of mature woodland and parkland, the Black Redstart prefers clefts in cliffs and rock-faces, from which he can dart out to catch passing insects, or hunt along the tide-line for small flies and crustaceans.

The odd thing is that he now shows a marked preference for industrial complexes, where he can find accommodation and food to his liking: wasteland and rubble-strewn areas provide plenty of insects.

The Black Redstart is much rarer than his cousin and until comparatively recently was only known as a winter visitor to the southwest of England. Its stronghold is central and southern Europe and the favoured wintering place is North Africa. It was not known to breed in Britain until 1923, when two pairs nested in Sussex between Hastings and Rye. Within a few years, it began colonising the London suburbs and, by 1933, a pair had even made its home in the heart of Woolwich Arsenal, surrounded by heavy traffic. In the following years, it moved into central London, and by 1940, a pair raised two broods in Westminster Abbey itself. Then came Hitler's blitz and with it the Black Redstart's greatest opportunity: plenty of ruins and open spaces covered in rubble which he put to excellent use. It was a curious bonus for the war-battered city of London.

Since then, with the rebuilding of London, the Firetails have made their homes in railway sidings, warehouses, gasworks and even nuclear power stations. In 1973 one of the, appropriately, soot and rust-coloured birds, was even heard giving his staccato warble from the inside of a jumbo-jet hangar at Heathrow.

Whatever Man may build or however he may rearrange the landscape, there is always some bird that will be able to make use of the new situation created.

BASIL EDE

Acknowledgements

Acknowledgement is gratefully made for permission to include extracts from the following works:

BB: *The Countryman's Bedside Book* (Eyre and Spottiswoode).

Davies, W.H.: *The Kingfisher* from *The Complete Poems of W.H. Davies* (Executors of the W.H. Davies estate).

De la Mare, Walter: *Martins:September* and *The Titmouse* from *The Complete Poems of Walter de la Mare* (The Literary Trustees of Walter de la Mare and the Society of Authors as their representative).

Harvey, F.W.: *Ducks* from *Ducks and other Poems* (Sidgwick and Jackson Ltd.).

Hodgson, Ralph: *The Missel Thrush* from *Collected Poems* (Mrs Hodgson and Macmillan, London and Basingstoke).

Home, Henry Douglas: *The Birdman* (William Collins, Sons & Company Ltd.).

Home, Lord: *Border Reflections* (William Collins, Sons & Company Ltd.).

Hughes, Ted: *Hawk Roosting* from *Lupercal* (Faber and Faber Ltd.).

Ivy, Robin: *Great Tit* and *Treecreeper* (By kind permission of the author).

Rittenhouse, Jessie B.: *Osprey and Eagle* from *The Secret Bird* (Houghton Mifflin Company).

Roethke, Theodore: *The Siskins* from *The Collected Poems of Theodore Roethke* (Faber and Faber Ltd.).

Rye, Anthony: *The Bullfinch, The Little Owl, Chiff-Chaff, The Lapwing, Titmice, Tree-creeper, The Wood Owl* from *The Inn of the Birds* (Jonathan Cape Ltd.).

Young, Francis Brett: *The Isle of Voices* from *The Island* (William Heinemann Ltd. and David Higham Associates Ltd.).

The publishers have made every effort to contact the copyright holders of the extracts used in this book. They apologise to any that have been inadvertently omitted and invite the copyright holders to contact them direct.